Endorsements

The prophet of values-centric living, Richard Barrett, has once again moved the goal posts for society by unveiling his sweeping analysis on the nature of evolution. Exploring Richard's book is a thoroughly enjoyable "ride." Through his incredible awareness he weaves a new narrative for conscious evolution, the benefits that flow from it, and the ways to achieve it. This book is a delight for those of us who have been fans of Richard's work for decades, and a great introduction to his amazing mind and heart for first time explorers of his very compelling worldview.

Rinaldo S. Brutoco, *Founding President and CEO, World Business Academy*

Yet again, Richard Barrett triumphs by picking our world up as if it were a snow globe, giving it timely shake and enabling us to make sense of what settles. He poses questions that cut right to the very heart of our current worldview. I was left with an even stronger urge to consider how I can be a greater contribution to our evolutionary development. If everyone read Richard's book, the world dynamic would quickly shift to a far happier, healthier and collaborative place.

Richard Tyler, *Chief Possibility Architect*

The Evolutionary Human is a must read at this critical juncture for humanity. It illuminates the fundamental attributes that make evolution possible and provides valuable new insights on how we can take our evolutionary journey in a more positive direction than generally anticipated.

Björn Larsson, *CEO, The ForeSight Group*

An important and succinct guide to evolving your personal and collaborative evolutionary intelligence in the context of a conscious, intelligent universe. This cosmology shows clearly how we can navigate current crises to reach our destined maturity as humans creating a world in which all living beings can thrive.

Elisabet Sahtouris, *PhD; evolution biologist and futurist, author of Gaia's Dance: The Story of Earth & Us*

Richard Barrett charts new territory in *The Evolutionary Human*, offering a fascinating and original theory of evolution that is richly supported by the evidence he marshals and explains in his compelling and highly readable style. There is nothing so practical as a good theory, and Richard shows the practical implications of his insights for how we can best govern our individual as well as collective lives. This is an important work from a modern master.

Raj Sisodia, *FW Olin Distinguished Professor of Global Business, Babson College; Co-founder and Co-Chairman, Conscious Capitalism Inc.*

Richard Barrett takes on nothing less than the theory of evolution and adds immeasurably to our understanding. The elegant simplicity with which Richard guides us through his ideas is refreshing in this era of complication and complexity and indicates a rare clarity of thought. His analysis of current world events and prognosis for the future should be required reading for all politicians and policymakers.

Ruth Steinholtz, *AretéWork LLP*

Richard Barrett is the leader of evolutionary consciousness of our times. Don't let the size of this small book fool you. Richard succinctly explains the new paradigm of evolution. Turns out evolution is intelligent and as a human family we need to tap into this intelligence not just to survive, but also to thrive and progress as humanity. Understanding how our evolution is designed gives us what we need to tackle the biggest global challenges to date and you will find an excellent framework for that in this book.

Bea Benkova, *Founder & CEO, and Tatiana Benkova, COO, Global Institute for Extraordinary Women, www.gifew.org*

In *The Evolutionary Human*, Richard Barrett presents an elegant theory of evolutionary intelligence in clear, compelling language. Affirming that we are all connected as aspects of a universal energy field, Barrett offers fresh insights into how the concept of survival of the fittest can be reinterpreted in light of the new physics and human psychological development. His explanation of how cooperation lies at the root of evolution across all levels offers great hope for the future.

Rev. Deborah Moldow, *Founder, The Garden of Light; Co-Director, The Evolutionary Leader's Circle of the Source of Synergy Foundation*

With his concept of Evolutionary Intelligence, Richard Barrett cracks the source code of human motivation. Whereas Charles Darwin found the recipe for the evolution of life forms, Richard Barrett goes further, he presents a unified theory of evolution that goes from the Big Bang to the present day. Absolutely brilliant.

Patrik Somers-Stephenson, *Facilitator of Transformation, Evolution Inside Out Ltd*

The Evolutionary Human is a profound and fresh narrative addressing the how and the why of our existence on Earth. Richard makes the invisible visible with simple language and scientific grounding. He offers a new invitation that considers consciousness the main reason of our human growth and development. This is groundbreaking! We can finally relax and allow life to flow.

Ina Gjikondi, *Director, Executive Education and Coaching, Center for Excellence in Public/ Leadership, George Washington University*

Richard Barrett has done it again. The "values maven" has brought his unique insights to Charles Darwin's concepts of evolution and the survival of the fittest. By comparing micro and macro systems, and stages of individual and collective psychological development, Barrett goes beyond neo-Darwinism and intelligent design to propose a whole new concept for understanding evolution. His eye-opening take on evolution will provide food for thought to political leaders, entrepreneurs and social change makers around the globe.

Marc J. Sachnoff, *Modern Wisdom Leadership Institute*

A bold thinker and savvy architect, Richard never stops to push the bound of human development theories and applications. This concise book provides a simple and accessible roadmap to personal growth. In this process, Richard presents a rare gift for the evolution of humanity!

Niran Jiang, *Co-founder and Director,*
Institute of Human Excellence

Once again Richard has created a book that is both practical and profound. As a clinical psychologist, coach, and designer of transformational events, I have been using Richard's tools and books with my clients to evolve their resilience and evolutionary consciousness for over 22 years. My clients love his useful maps for the practical navigation of the visible and invisible worlds. With each of his books, Richard manages to go deeper.

Erik Muten, *MFA, PsyD, Clinical Psychologist, Mentor,*
Organizational Consultant and Designer of Transformational
Events, Kailo Mentoring Group/DramaWorks Interactive

The historians, philosophers, and psychologists, who have studied human evolution all agree: for some reason humanity evolves not continuously but by sudden leaps. The book, *The Evolutionary Human* by Richard Barrett, is about to make a new leap in the theory of evolution.

Miša Lukić, *Founder and Chief Business Designer, New Startegy*

The Evolutionary Human is incisive in its style and holistic in its grip: cutting through academic swaths and psychological complexities which often bog down other pioneers. The long and the short of it is we are re-awakening to our deeper and truer purpose as Homo sapiens and in so-doing we are remembering the wisdom that is innate within nature. Barrett's work is an important aid in this remembering and recognizing of our evolutionary potential beyond the self-limiting belief of the survival of the fittest into a multi-level evolutionary theory that provides for a richer understanding of human individuation and collective transformation.

Giles Hutchins, *Chair of The Future Fit Leadership Academy, and author of the books The Illusion of Separation and Future Fit*

Books by Richard Barrett

Worldview Dynamics and the Consciousness of Nations (2019)

Everything I Have Learned About Values (2018)

The Values-Driven Organization: Cultural Health and Employee
Well-being as a Pathway to Sustainable Performance (2017)

A New Psychology of Human Well-Being: An Exploration of the
influence of Ego-Soul Dynamics on Mental and Physical Health (2016)

The Metrics of Human Consciousness (2015)

Evolutionary Coaching: A Values-based Approach
to Unleashing Human Potential (2014)

The Values-Driven Organisation: Unleashing Human
Potential for Performance and Profit (2013)

What My Soul Told Me: A Practical Guide to Soul Activation (2012)

Love, Fear and the Destiny of Nations: The Impact of the
Evolution of Human Consciousness on World Affairs (2011)

The New Leadership Paradigm (2010)

Building a Values-Driven Organization: A Whole System
Approach to Cultural Transformation (2006)

Liberating the Corporate Soul: Building a Visionary Organization (1998)

A Guide to Liberating Your Soul (1995)

THE
EVOLUTIONARY
HUMAN
HOW DARWIN GOT IT WRONG

RICHARD BARRETT

"It was never about species.
It was always about consciousness."

Lulu Publishing Services rev. date: 12/11/2018

DEDICATION

To my wife, Christa, a loving, constant support, who somehow puts up with my strangeness, keeps me grounded, and makes sure the ideas I express in my books are at least halfway sensible.

TABLE OF CONTENTS

LIST OF FIGURES

LIST OF TABLES

FOREWORD

MY PURPOSE IN WRITING THIS book is threefold:

a. to challenge the current evolutionary paradigms—Neo-Darwinism and Intelligent Design—by providing a third, more holistic paradigm, based on evolutionary intelligence;
b. to explain how evolutionary intelligence operates; and
c. to explore the implications of this more holistic paradigm on the evolution of the species we call *Homo sapiens*.

My basic hypothesis is that evolution was never about species, it was always about consciousness. Therefore, the continuance of 3.8 billion years of evolution on Earth now rests on the shoulders of the evolution of human consciousness. Unlike Neo-Darwinism and Intelligent Design, which are only concerned with life on Earth, the evolutionary intelligence theory explains the whole of evolution from the Big Bang to the present day.

At the heart of this new theory is the scientific understanding that we live in an energetic world, and that our material world is a property of our limited senses. We view the world not as it is, but as we are: we are only able to perceive a small band of energetic frequencies which we interpret as our material reality.

The energetic theory of evolution postulates that there are four planes of being—the energetic plane, the atomic plane, the cellular plane, and the plane of creatures. Evolution progressed by entities at each plane of being creating a stable energetic platform from which the next plane of being could develop. Each stable energetic platform was created in three stages: first, individual entities learned how to become viable and independent in the energetic framework of their existence; then, these entities learned how to bond to form viable, independent energetic group structures; and finally, these viable, independent energetic group structures learned how to cooperate to create a higher order entity that became the stable energetic platform for the emergence of the next plane of being.

The only entity able to create a sufficiently stable energetic platform at the atomic plane was the carbon atom. The only entity able to create a sufficiently stable energetic platform at the cellular plane was the eukaryotic cell. And, potentially, the only entity that appears to be able to create a sufficiently stable energetic platform at the plane of creatures is *Homo sapiens*. (Fanfare of trumpets) Welcome to the evolutionary human.

PART 1

UNDERSTANDING EVOLUTION

1

WHAT IS EVOLUTIONARY INTELLIGENCE?

SIMPLY STATED, EVOLUTIONARY INTELLIGENCE IS *the ability of an entity to continuously adapt to changes in its environment, so it can thrive and prosper.* To do this, an entity must be able to do four things:

1. **Gather information:** The entity must be conscious of the changes that are occurring in its environment: it must be able to sense changes.
2. **Process information:** The entity must be able to draw together the different strands of information perceived by its sensing mechanisms to create an overall picture of the changes that are occurring in its environment.
3. **Make meaning of the information:** The entity must be able to analyze how the overall picture of the changes that are occurring affect its ability to maintain its internal stability and external equilibrium—enable it to meet its most important (survival, safety and security) needs.
4. **Decide what to do:** The entity must be able to choose a response or action to the changes that are occurring that allows it to maintain (or regain) its internal stability and external equilibrium.

If an entity is unable to maintain its internal stability and external equilibrium it will not be able to thrive and prosper; it will struggle to function and will eventually disintegrate or decompose into its component parts. This is true for all entities, from atoms, to molecules, to cells, to organisms, to creatures, including *Homo sapiens* and all human cultural/societal group structures.

Two types of adaptation

If we look at evolution from the Big Bang to the present day we notice two types of adaptation: physical adaptation and mental (psychological) adaption. As far as life on Earth is concerned, prior to the appearance of *Homo sapiens,* these two types of adaptation progressed, more or less, in parallel. Physical adaptation led to species evolution, and species evolution was usually accompanied by psychological evolution—an expansion of conscious awareness and intelligence.

However, once *Homo sapiens* arrived on the evolutionary scene, evolution stopped being about physical adaptation and became all about psychological adaptation—personal psychological evolution and collective psychological evolution. Personal psychological evolution led to new stages of collective psychological evolution, and new stages of collective psychological evolution generated changes in the human cultural/societal environment that fostered new stages of personal psychological evolution.[1]

Unlike most books on evolution, the focus of this book is on psychological evolution; not just human psychological evolution but the evolution of awareness of all entities in the chain of evolution from the Big Bang to the present day.

You may have heard the term "psychological evolution" used in the context of the stages of human development, but you may not have heard it used in the context of our material world—the world of particles, atoms, and molecules, nor in the context of cells and multicellular organisms. What I am suggesting, by using the term "psychological evolution" in this more holistic way, is that all entities that make up our material world have conscious awareness—they gather information, process that information, make meaning of that information, and decide how to respond or react to changes (stimuli) in a manner that gives them the best chance to maintain

---

[1] In my other works, I refer to the stages of collective psychological evolution as worldviews. I postulate that each new worldview (stage of collective psychological development) provides the contextual framework for human individuals to meet the needs of the next stage of their personal psychological development.

their energetic stability. Their ability to do this depends on achieving a correspondence between the level of functional/operational complexity of their minds and the level of complexity of their environments.

Thus, as evolution progressed—from elementary particles to protons, neutrons, and electrons, to atoms, to molecules, to cells, to multicellular organisms, to creatures, to human beings and their cultural and societal constructs—the complexity of the functioning of the minds of each of these entities was more or less "forced" to increase in parallel with the level of complexity of their environments if they were to meet their survival, safety and security needs in our three-dimensional material world.

Researchers, Michael Commons and Francis Richards suggest there are sixteen stages of hierarchical mind complexity starting at the molecular level—Stage 0. Cells operate at Stage 1, a new born human baby operates at Stage 3, and a 7–11 year-old operates at Stage 9. Stages 10–16 can only be found in humans.[2] My theory of evolutionary intelligence starts much earlier, at the level of electronic particles. In other words, at the level of the basic building blocks of our material world.

Consequently, because of this hierarchy of mind complexity, the mind of a cell is more complex than the mind of a molecule. Why? Because a cell lives in a more complex environment than a molecule, and if the molecules that make up a cell cannot maintain their internal stability and external equilibrium, the cell cannot maintain its stability.

Furthermore, if the cells that make up a creature or a human being cannot maintain their internal stability and external equilibrium, the creature or human cannot maintain its stability. Similarly, if an individual human cannot maintain its internal stability and external equilibrium, the family it is part of will have difficulty maintaining its stability, and so on; if communities cannot maintain their internal stability and external equilibrium, then nations cannot maintain their stability. In other words, the whole of our human experience depends on the ability of all the entities we depend on for our existence at each lower plane of being (level of existence) maintaining their internal stability and external equilibrium—living in harmony with themselves and their environment.

---

[2] https://en.wikipedia.org/wiki/Model_of_hierarchical_complexity#Stages_of_development

The fundamental role of decision-making

When faced with a change in its environment, an entity must first consider if the change is threatening—potentially destabilizing—or not. If it judges, based on memories of past experiences, that the change is not potentially destabilizing, it does nothing. If it judges, based on past experiences, that the change is potentially destabilizing, it must decide how to respond or react to preserve or regain its stability. If it has experienced a similar change previously and found a solution to maintaining its stability, it will use that solution again. This is called an instinctive response. However, if it is a new threat, one it has never experienced before, evolutionary intelligence is triggered.

Evolutionary intelligence contains three decision-making algorithms.[3] The algorithm chosen by the entity depends on two factors: the level of psychological development of the entity, and its physiological and psychological ability to engage with the second and third algorithms of evolutionary intelligence—to bond with other entities and/or for the group structure it is part of to cooperate with other group structures. The three algorithms and the questions that dictate which algorithm an entity uses are as follows:

1. **Becoming viable and independent:** Can I (the entity) overcome this threat by becoming stronger or more resilient and thereby return to stability?
2. **Bonding to form a group structure:** Can I (the entity) bond with another entity that is facing the same threat, to form a temporary or permanent group structure that is strong enough or resilient enough to overcome the threat so both of us can return to stability?
3. **Cooperating to form a higher order group structure:** Can we (the entities that are part of a group structure) cooperate with another group structure in a temporary or permanent higher order group structure that is strong or resilient enough to overcome the threat, so we (all group structures and all entities contained therein) can return to stability inside the new, higher order group structure?

---

[3] See the appendix for an explanation of this term.

Viewed from this perspective we can make the following statements:

1. The purpose of conscious awareness is always the same—to enable an entity to maintain its internal stability and external equilibrium by finding ways to overcome threats to its existence and/or meet its most important needs;
2. Evolutionary intelligence is a property of consciousness;
3. Consciousness and evolutionary intelligence are fundamental characteristics of all entities found in our three-dimensional material world. Not only is the world we live in energetic and conscious, it is also intelligent.
4. Evolutionary intelligence gives purpose and direction to evolution.

So, as humans, whenever we encounter a threating situation we have never experienced before, or a situation in which we find it difficult to get our needs or desires met, evolutionary intelligence steps in to provide a way for us to overcome the threat and/or get our needs or desires met.

In other words, evolutionary intelligence is the source code of human motivation. Every single decision we make, be it a conscious,[4] subconscious,[5] or unconscious,[6] has the same objective—to get our most important needs or desires met so that we can maintain our energetic stability. Evolutionary intelligence allows us to satisfy:

- the needs of the body, so we can stay alive;
- the needs of the ego,[7] so we can survive, keep safe, and feel secure in our physical and cultural framework of existence—our deficiency needs; and
- the desires of the soul, so we can find fulfilment in this lifetime— our growth needs.

Your ego cannot meet its survival, safety, and security needs if your body cannot stay alive, and your soul cannot satisfy its desires if your ego

---

[4] Ibid.
[5] Ibid.
[6] Ibid.
[7] Ibid.

cannot create the conditions that allow you to survive, keep safe, and feel secure. Only when the body is in good health, and we have mastered our deficiency needs (survival, safety, and security needs), are we able to focus on mastering our growth needs—our soul's desire for fulfilment—self-expression, connection, and contribution.

In the words of Abraham Maslow: "Our deficiency needs are prepotent to our growth needs. We get anxious and fearful if our deficiency needs are not met, but when we *believe* they have been met, we give them no further consideration."

I have italicized the word *"believe,"* because it doesn't matter how rich you are, how loved you are, or how much you are held in the esteem of others, if you don't believe you have enough, you don't believe you are loved enough, or you don't believe you are enough, then you will not consider your deficiency needs have been met. You will still be anxious and fearful, and you will never be able to give your full attention to your growth needs.

There is hardly any activity we undertake that does not involve trying to satisfy our body's needs, our ego's needs, or our soul's desires: sleeping, keeping fit, achieving goals, relaxing to music, singing in a choir, raising kids, writing a book—they are all attempts to satisfy a body need, an ego need, or a soul desire.

Even when we are helping others to get *their* needs met, we are attempting to satisfy either our ego's need to find meaning and purpose (our soul's desire for self-expression), our ego's need to make a difference (our soul's desire for connection), our ego's need to be of service to others (our soul's desire for contribution), or alternatively our ego is hoping that by helping others to get their needs met, the beneficiaries of our actions will reciprocate at some point in the future and help us to get our needs met.

Every time we fail to meet our body's or ego's needs or our soul's desires, we experience psychological instability. And, every time we experience psychological instability, our evolutionary intelligence kicks in to try to help us regain our energetic stability by helping us to get our needs or desires met. If we are unable to get our needs or desires met, we experience either instability in our body (physical dysfunction) or instability in our mind (mental dysfunction). The whole of human history can be explained by our attempts to satisfy our body's or ego's needs or our soul's desires.

2

BEYOND DARWINISM

IF YOU HAVE HAD A classical education in science, when you think of the word "evolution" you will immediately think about Charles Darwin and the evolution of species. This is because modern science views our world through the lens of material awareness; it does not view our world through the lens of energetic awareness.

Viewed through the lens of energetic awareness, evolution has always been about consciousness, it was never about species. The fossils that our material scientists study, and the life forms they dissect, are simply the remains of vehicles of consciousness. Viewed from the more holistic energetic perspective, the arrow of evolution has always been about developing the capacity of "mind" to increase its sense of inclusion and decrease its sense of separation in increasingly complex frameworks of existence.

Our scientists rarely speak about this because, for them, the topics of mind and consciousness are off limits. They rarely venture into the realms of consciousness and the world of psychology. Despite the discovery of the quantum field over sixty years ago, our material scientists have not been unable to accept the idea that the world we live in is an energetic continuum.

The Big Bang

According to modern scientific theories, everything that exists in our universe originated from an enormous energetic explosion that occurred about 13.8 billion years ago. From that point on, it was all about evolution: the evolution of energy into matter, matter into living organisms, and living organisms into creatures. One of those creatures—*Homo sapiens*—is now attempting to carry the baton of evolution to the next level by making the concept of humanity palpable.

When I say that everything in the universe had its origins 13.8 billion years ago, I literally mean everything, including not only the physical world of atoms, cells, and *Homo sapiens,* but also the energetic world of instincts,[8] thoughts, feelings, beliefs,[9] and values. Indeed, evolution would not have occurred if the faculties we attribute to our senses and the physical brain (data gathering and information processing) had not evolved in parallel with the faculties we attribute to the mind (meaning-making and decision-making).

If the Big Bang theory is correct, then it follows that the physical world emerged from the energetic world. Not only did energy precede matter, we know, thanks to Einstein, that energy and matter are related ($E=mc^2$): energy is equivalent to matter times the speed of light squared. In other words, energy is the fundamental backdrop to our physical universe.

Consequently, we find ourselves living in two worlds: the three-dimensional (3-D) material world of the body (the tangible part of our existence) and the (4-D) energetic world of the mind (the intangible part of our existence). Although mystics and shamans have been aware of the underlying unity of our material and energetic worlds for millennia, it wasn't until the early part of the twentieth century and the discovery of the quantum energy field that scientists began to acknowledge that there was a crack in our material interpretation of reality.

Ervin László, a Hungarian-born philosopher of science, describes the two-world problem in the following way: he calls the observable, manifest, material 3-D world the M-dimension (M for manifest), and he calls the unobservable, energetic 4-D world the A-dimension. The A-dimension (A for Akashic or energetic dimension) is a universal field of information and potentiality that is in constant interaction with the M-dimension.

> The A-dimension [energetic dimension] is prior: it is the generative ground of the particles and systems of particles that emerge in the M-dimension [material dimension].[10]

---

[8] Ibid.

[9] Ibid.

[10] Ervin László, *The Self-Actualizing Cosmos: The Akasha Revolution in Science and Human Consciousness* (Rochester: Inner Traditions), 2014.

Max Planck (1858–1947), a theoretical physicist, who was one of the originators of quantum theory, is quoted as saying: "I regard consciousness as fundamental. I regard matter as derivative from consciousness. We cannot get behind consciousness."

Pioneering physicist, Sir James Jeans (1877–1946), expressed a similar thought when he wrote:

> The stream of knowledge is heading toward a non-mechanical reality; the Universe begins to look more like a great thought than like a great machine. Mind no longer appears to be an accidental intruder into the realm of matter ... we ought rather hail it as the creator and governor of the realm of matter.[11]

The physicist, Richard C. Henry, is more adamant. He says:

> Get over it, and accept the inarguable conclusion. The Universe is immaterial—mental and spiritual.[12]

To get a deeper understanding of the difference between a worldview based on 3-D material reality and a worldview based on 4-D energetic reality, I am going to ask you to participate in an exercise.

The five-finger exercise

Take a flat sheet of paper and imagine that there is a small person living on the surface of the paper in what is known as "Flatland." For this person, the world has length and breadth but no height. In other words, this person operates in a world of two-dimensional awareness (2-D awareness): she cannot perceive height. Now along comes a human being (you) with 3-D awareness (you can perceive height), and you place the fingers of one hand on the paper on the surface of Flatland (see figure 2.1).

---

[11] R. C. Henry, "The Mental Universe," *Nature* 436: 29, 2005.
[12] Ibid.

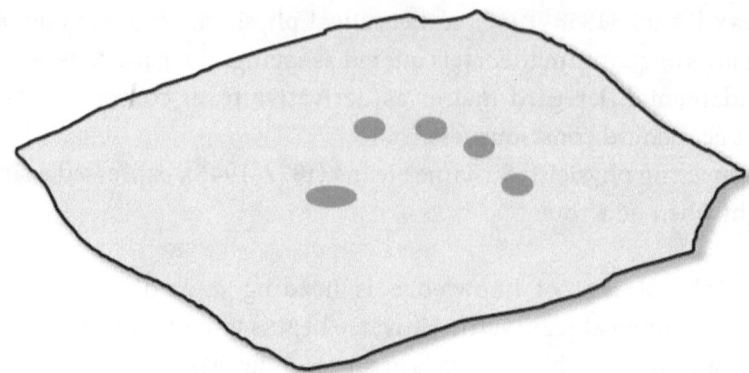

Figure 2.1: The projection of five fingers into two-dimensional awareness

Imagine now that the person living in Flatland is out for a morning stroll. When passing this place yesterday, she noticed nothing unusual. Suddenly, overnight, five circles have appeared (the projection into two-dimensional awareness of your five finger tips). The two-dimensional being is mystified by the appearance of the five circles. She calls her friend, a two-dimensional scientist, and asks her to explain the nature of the five circles. The two-dimensional scientist explores the five circles using her two-dimensional logic.

Her experiments show that the circles can move independently within certain limits, but if she puts a large enough force on one circle, eventually it will appear to drag the other circles with it (although the fingers of the hand are separate, they are connected, but in a dimension of awareness that the two-dimensional scientist cannot perceive).

The two-dimensional scientist repeats her experiments. She builds sets of equations to explain the strange relationship of the circles to one another, and, before too long, she believes she knows everything there is to know about the five circles, except where the circles came from. The two-dimensional scientist publishes a paper about the circles and calls a meeting of the Academy of Two-Dimensional Scientists to show them her discovery. Several other two-dimensional scientists repeat the experiments and get very similar results. Everyone in the two-dimensional world believes they know all there is to know about the five circles, except their origin. The origin of the circles becomes a focus of speculation and intense debate among the scientists and religious clerics who inhabit Flatland.

Viewed from our perspective of 3-D awareness, we know that these are not five separate circles. They represent five connected fingers that form part of a living organism. The two-dimensional beings are completely unaware of this larger picture. They believe the five circles are separate but linked in some manner that they cannot quite understand. They have equations to explain these linkages, but their equations are symbolic: they do not convey the full reality of the connection that exists at the third dimension of consciousness. This is because the two-dimensional scientists are tuned in to the frequencies of two-dimensional awareness and cannot pick up the frequencies of three-dimensional awareness.

This is exactly the situation we find ourselves in with regard to the fourth dimension of consciousness. We have countless experiences that appear physically unconnected but, in reality, are causally linked. They are linked in the fourth dimension of consciousness. Some of these experiences we can explain with our 3-D logic—this is the domain of science—and some we cannot explain. We have a range of names for the experiences we cannot explain. Depending on the context, we can call them paranormal, magical, miraculous, spiritual, or synchronistic. These are simply words we use to disguise our ignorance.

Because we rely on our 3-D awareness, we cannot perceive the linkages that exist in the higher dimensions of consciousness from which these experiences originate. The reason they are beyond our comprehension is that our minds are focused on the partial information we receive from our five physical senses. Just as in Flatland, where people's awareness was tuned in to 2-D awareness, our physical senses are only tuned to in to 3-D awareness—our material reality. Therefore, as long as we identify with our physical body and its senses, we will be unaware of what is happening in our 4-D energetic reality.

This statement by scientist Bernardo Kastrup explains the situation:

> The function of the brain is to localize consciousness, pinning it to the space-time reference point implied by the physical body. ... When not subject to this localization mechanism, the mind is unbound. ... Therefore, by localizing the mind, the brain 'filters out' of consciousness anything that is not correlated with the body's perspective ... like a radio receiver selecting from among [a] variety of stations.[13]

---

[13] Bernardo Kastrup, *Why Materialism Is Baloney* (Winchester: Iff Books), 2014, p. 40.

What we can conclude from the five-finger exercise is that materiality is not a property of the world; it is a property of our senses. What we observe in our 3-D world is energy in the form of matter. Edward O. Wilson, Professor Emeritus at Harvard University and one of the world's pre-eminent biologists and naturalists, agrees on this point. He states:

> We are aware of only minute slivers of space-time, and even less of the energy fields, in which we exist.[14]

The comb analogy

There is another way of illustrating the two-world problem. Take out a comb and cover up the top half. What you see are the unconnected teeth of the comb. You see the separation. When you uncover the top half of the comb, you see that the teeth are joined at a higher level. You see connectedness; you see unity. Without the higher-level connection, the comb would fall apart and would not fulfil a useful purpose. The connection gives the separate teeth a larger meaning.

As human beings, this is how we are: what we perceive through our 3-D physical senses are separate human beings (the teeth of the comb). When we raise our awareness to the energetic fourth-dimension of consciousness, we can begin to recognize and understand our connection to others; we begin to appreciate the unity.

Just as it is difficult to understand the meaning and purpose of the separate teeth until we are aware that they belong to a comb, so too it is difficult for us to understand our meaning and purpose until we become aware of our connection to other people. We are not separate; we are all connected—each one of us is an individuated aspect of the same universal energy field.[15]

As long as we identify with the body and rely only on our physical senses to interpret reality, what we can see, hear, smell, taste, and touch are the objects and sounds that vibrate at frequencies that our senses can detect. The rest of the energetic world goes unnoticed. The way we connect

---

[14] E. O. Wilson, *The Meaning of Human Existence* (New York: Liveright Publishing), 2014, p. 166.
[15] See the appendix for an explanation of this term.

with this larger world is through our feelings, our intuition, synchronicity, and inspiration. We also connect with the 4-D world through clairvoyance (the ability to perceive future events), clairaudience (the ability to hear voices of non-sentient beings), clairsentience (the ability to read the feelings of others), and claircognizance (the ability to receive knowing insights as well as paranormal experiences—events and experiences that have no scientific explanation). In the words of Leonard Cohen, "There is a crack in everything. That's how the light gets in."[16]

Newtonian mechanics versus quantum mechanics

Based on this understanding, it becomes obvious that if we rely solely on our 3-D awareness to explain our reality, we will not have a complete idea of who we are or the world in which we live. Just like the two-dimensional scientists who believed they knew all there was to know about the five separate circles, until about a century ago our three-dimensional scientists believed they knew all there was to know about our material world. Everything could be explained by Newtonian mechanics.

In 1900, Lord Kelvin (1824–1907), a British mathematical physicist and engineer, summed up the state of science in the following statement: "There is nothing new to be discovered in physics now. All that remains is more and more precise measurement."

Twenty-five years later, quantum theory began to emerge onto the scientific stage. Our three-dimensional scientists had to face up to the fact that there was a crack in 3-D material reality. They discovered that there was a macro world of objects (made up of particles and atoms), that we find in our 3-D material world, and a micro world of energy (waves of information), that we find in our 4-D energetic world.

What we discovered through quantum theory was that all that was necessary to bridge these two worlds was an observer. When a person (observer) operating from 3-D material awareness is present, the energy waves collapse into a specific outcome that aligns with the beliefs of that person. Two people can interpret the same event in different ways, and, based on their beliefs, they will have different emotional experiences.

---

[16] The song called "The Anthem."
https://www.youtube.com/watch?v=mDTph7mer3I

In other words, our personal beliefs and our worldviews can lead us to different interpretations of reality. We see the world as we are. Through the discovery of quantum theory, physicists were able to connect the causal energy of our 4-D mind (beliefs) with the outcomes we experience in our 3-D material reality.

The reason we are so mesmerized by classical Newtonian mechanics is that it accurately describes the movement of objects and systems susceptible to our 3-D human perception—objects that are larger than a molecule and smaller than a planet, at temperatures close to those that living organisms can survive in and going at speeds significantly less than the speed of light. When these boundary conditions are transcended, and we move into the micro world of atomic particles, classical mechanics no longer holds sway, and quantum mechanics takes over. Everything in the universe has an energetic dimension, including every aspect of our physical bodies. Even the brain, our DNA,[17] and our physical senses have energetic dimensions. In other words,

> [If you focus] on the structure of the atom, you would see nothing, you would observe a physical void. The atom has no physical structure, we have no physical structure. ... Atoms are made out of invisible energy, not tangible matter.[18]

Niels Bohr (1885–1962), one of the originators of quantum theory, puts it like this: "Everything we call real is made of things that cannot be regarded as real."

What is "real" is not what we see, hear, smell, taste, or touch; it is what we feel. Everything we see, hear, smell, taste, or touch is a 3-D material representation of a 4-D energy field. Our emotions and our feelings are our connections to that energetic world. What we feel are energy shifts: energy in motion—our emotions. Feelings are the cognitive interpretation of our emotions, which are conditioned by our beliefs. We cannot know what is happening in our energy field if we are not aware of our feelings.

---

[17] See the appendix for an explanation of this term.
[18] Arjun Walia, "Nothing Is Solid & Everything Is Energy—Scientists Explain the World of Quantum Physics," *Collective Evolution*, September 27, 2014.

For the most part, evolutionary biologists view the world through their 3-D material perspective. They are not aware of the energetic dimension of our existence, or, if they are, they choose to ignore it. For them, consciousness is an epiphenomenon of the brain, and without a brain there is no consciousness. Just like the two-dimensional scientists I spoke of earlier, they cannot see beyond their 3-D material world. This is why they regard the topic of the mind and consciousness as off limits and why they find it difficult to find an explanation for any form of evolution other than Darwinism.

Evolution 1.0

Charles Robert Darwin (1809–1882) was one of the first to explore evolution in a scientific manner. His theories started to become vogue around 1859, when he published his treatise *On the Origin of Species by Means of Natural Selection*.[19] Darwin introduced the idea that species evolve over the course of generations though a process of random mutation and natural selection. For the moment, let's call this Evolution 1.0. We can express Darwin's theory by the following formula:

Random Mutation + Natural Selection + Time = Evolution 1.0

In this formula, "random" means proceeding without a definite aim, reason, or pattern. "Mutation" means a departure from the parent type in one or more heritable characteristics, caused by a random change in a gene or a chromosome.[20] "Natural selection" means a process that results in the survival and reproductive success of individuals or groups best adjusted to their environment that leads to the perpetuation of genetic qualities best suited to that environment. In Darwin's theory, the focus is on physical adaptation. There is very little, if any, focus on psychological adaptation.

---

[19] Charles Darwin, *On the Origin of Species by Means of Natural Selection* (New York: D. Appleton and Company), 1859.
[20] When Darwin put forward his theory of evolution, genes and chromosomes had not been discovered. The discovery of genes in the middle of the twentieth century confirmed for many the legitimacy of Darwin's concepts.

Creationism

Darwin's ideas about evolution were immediately controversial, because he challenged the generally accepted natural theology of the Church of England. The idea that one species could transmute into another was antithetical to the Church. The Church believed that species were part of an unchanging natural hierarchy and that humans were not only unique but unrelated to other animals. These ideas became known as Creationism.

Creationism is the religious belief that the universe and life originated from a divine act rather than natural processes. Creationists base their belief on a literal reading of Christian and Islamic religious texts.

Within about two decades of the publication of *On the Origin of Species,* there was widespread scientific agreement that evolution, with a branching pattern of common descent, could explain the diversity of life we find on Earth. By the 1940s, less than a century later, Darwin's concept of evolutionary adaptation through random mutation and natural selection had become an accepted scientific fact. For those who were still doubtful, the discovery of DNA confirmed his thesis—*Homo sapiens* is a product of evolution.

Intelligent Design

Meanwhile, the idea of creationism evolved into what is known as the Theory of Intelligent Design. The proponents of Intelligent Design believe that certain features of living things are best explained by an intelligent cause not an undirected process such as random mutations and natural selection. They believe that the randomness of Darwinism implies that humans have no spiritual nature, no moral purpose, and no intrinsic meaning. Creationists seek to reverse the dominance of the materialistic worldview through the acceptance of a theistic worldview.

Neo-Darwinism

From the middle of the twentieth century onwards, new evidence from the fields of genetics, systematics, and palaeontology established Darwinian evolution as biology's central paradigm. Genetics is the study of genes, genetic variation, and hereditary. Systematics is the study of the evolutionary relationship between living forms over time. Palaeontology is the study of fossils to determine an organism's evolution and relationship with its environment. The adherents to these theories are known as Neo-Darwinists.

This is where we find ourselves today: two competing evolutionary theories whose proponents spend their time savagely attacking each other—on the one hand, the Neo-Darwinists and on the other, the proponents of Intelligent Design. Each group is trying to prove that they are right and that the other group is wrong. The epicentre of this war is in the US, where the proponents of Intelligent Design want to see their theory taught in schools as an alternative to Neo-Darwinism.

Evolution 2.0

In *Evolution 2.0: Breaking the Deadlock between Darwin and Design*,[21] engineer Perry Marshall makes a bold attempt to move the debate forward by trying to convince us that both sides are right, and both sides are wrong. I find myself in alignment with many of his ideas, not only because I find his arguments rational and convincing but also because they provide a springboard for my own ideas about evolution, which I will describe shortly and refer to as Evolution 3.0.[22]

Marshall states: "In one corner we have Neo-Darwinian, atheists like Richard Dawkins, Daniel Dennett and Jerry Coyne who insist evolution happens by blind random accident, and in the other corner we have

---

[21] Perry Marshall, *Evolution 2.0: Breaking the Deadlock between Darwin and Design* (Dallas: Benbella Books), 2015.

[22] I think it is important to recognize at this point that the terminology I am using could be a little misleading for some people. Let me explain. The use of terms such as "1.0," "2.0," and "3.0" is normally reserved for more advanced versions of a software program. If you have a smart phone, you will be familiar with this concept; if you use Apple technology, you will be constantly receiving program updates such as iOS 6.5.1, etc. These are operating system upgrades. I am not using the terms Evolution 1.0, 2.0, and 3.0 in the same manner. I am using them to refer to upgrades in our understanding of the process of evolution.

Intelligent Design advocates like William Dembski,[23] Stephen Meyer and Michael Behe who maintain that evolution is a fraud, rejecting common ancestry outright."

Marshall believes that his third way, Evolution 2.0, "proves that, while evolution is not a hoax, neither is it random nor accidental": the changes that living entities make as their life conditions shift are adaptive, goal orientated, and conscious. In other words: evolution is purposeful. Marshall refers to purpose-driven evolution as Evolution 2.0. He defines it in the following way:

Adaptive Mutation + Natural Selection + Time = Evolution 2.0

The concept of adaptive mutation is potentially a major game changer in our understanding of evolution because it suggests purpose and intelligence. It also hints at direction. Whereas random mutations produce random outcomes, adaptive mutations set out to achieve a goal. In other words, there is a purpose behind the so-called mutation.

Marshall suggests that over time the accumulation of purposeful outcomes provides a direction to evolution. This poses the question: "What is this purpose"—this goal that drives adaptive mutations? The answer is as simple as it is obvious. I believe the purpose of adaptation is to enable entities to stay alive—stay present in 3-D material awareness. In other words, adaptation is driven by the will to live. We call the result of thousands of such adaptations "evolution." Those entities that achieve the goal of staying alive; those that are able to adapt or overcome existential threats, provide the genetic material for natural selection. The species that can't adapt in the face of life challenges—perish and become extinct.

Evolution 3.0

I believe that to achieve any goal, particularly a complex goal such as staying alive, requires thought and intelligence, along with experimentation. This means we can replace the term "natural selection" with the term

---

[23] On September 23, 2016, Dembski announced his official retirement from Intelligent Design, resigning all his formal associations with the Intelligent Design community.

"intelligence," and we can replace the term "adaptive mutation" with the term "adaptive thinking." Thus, we can rewrite the formula for evolution in the following way:

Adaptive Thinking + Intelligence + Time = Evolution 3.0

Bringing together intelligence with adaptive thinking, we arrive at the concept of evolutionary intelligence. Thus, we can state:

Evolutionary Intelligence + Time = Evolution 3.0

If we want this formula to be accepted by the scientific community, we must define what evolutionary intelligence means. Why? Because most scientists are unable to accept the idea that life forms without a brain, such as cells, are intelligent and capable of adaptive thinking. This is a ridiculous assumption that is easily refutable. For example, it is common knowledge that new strains of virus are constantly appearing as they learn to overcome the threat that new drugs have on their survival. They don't need a brain to evolve, they just need a mind.

Every living entity has a mind, including the cells in our body, and every mind is primarily focused on staying alive and procreating. Enabling an entity to stay alive is the fundamental purpose of evolutionary intelligence. If an entity cannot stay alive, it cannot procreate. Without the *will to live and procreate,* evolution would never have happened. The Earth would be a dead planet. This raises an important question: "Where does evolutionary intelligence—the intelligence that is required to stay alive and support procreation (keep the 'body' functioning and the species alive)—come from?"

The answer that most 3-D material scientists would give to this question is DNA. Viewed through the lens of 3-D material awareness, deoxyribonucleic acid (DNA), a molecule that contains the genetic instructions used in the development, functioning, and reproduction of living organisms, contains all the instructions necessary for an entity to stay alive, procreate, and grow. But does it contain the instructions that trigger the will to survive? I believe the answer is no!

For life to flourish, there must be a psychological driving force that activates the instructions contained in DNA for the growth and development of an organism and trigger its homeostatic functioning.

Let me provide a metaphor to explain what I mean. You decide you need a new desk for your home office. You go to the DIY store and buy a self-assembly kit. The kit comes in components with instructions about how to assemble the desk. *It does not come with the will to assemble the desk.* The will to assemble the desk comes from an outside source: you! You supply the will to assemble the desk. In other words, the components and instructions are separate from the will to assemble the desk. The instructions and components come from one dimension of reality—the material and physical dimension, and the will to assemble the desk comes from another dimension of reality—a mind: your mind.

DNA is very similar. DNA is a set of instructions for assembling components that allow an entity to stay alive, grow, and procreate. Each creature and living entity has their own set of unique instructions. The development of the DNA molecule has been going on for more than three billion years—from the moment the first spark of "life" appeared on Earth. Over time the instructions in DNA got modified, resulting in the plethora of life forms (species) we find on Earth.

What is common to all the different life forms is a set of instructions for staying alive and procreating—staying present in 3-D material awareness. What you do not find in the DNA of any life form are a string of genes that trigger the will to live. That is because the will to live is supplied by an external force, and this external force is also the source of the conscious awareness of the life form.

I have searched everywhere to find an answer to the question: Where does the will to live come from? The closest answer I have found is an answer to the question: "What *is* the will to stay alive?"

"The will to live [stay alive] is a psychological force [causing entities] to fight for self-preservation which is seen as an important and active process of conscious and unconscious reasoning [motivation]."[24] This psychological force (motivation) plays a significant role in the decision-making of all sentient beings.

I conclude, therefore, that *the will to live or stay alive* is an external psychological force rather than a biological force, because it cannot be found anywhere in DNA. Furthermore, staying alive until maturity is important because it allows procreation to take place. In other words, the will to live and stay alive and the will to procreate have the same objective as

---

[24] https://en.wikipedia.org/wiki/Will_to_live

far as the external psychological force is concerned: they are both necessary for a species to stay present in our 3-D material world.

I conclude that without these psychological imperatives (staying alive and procreating) and access to evolutionary intelligence, life would not have evolved on Earth. Thus, we can rewrite the formula for evolution in the following way:

Energetic Instability (fear) + Evolutionary Intelligence + Time = Evolution 3.0

Whenever the conscious awareness of an entity experiences a threat to meeting its needs that it has not experienced before, evolutionary intelligence is triggered. In other words, it is the fear in the mind of an entity about staying alive or meeting its needs that triggers evolutionary intelligence. Fear at the level 4-D awareness represents energetic instability. When an entity lives in harmony with its surroundings the mind of the entity is energetically stable. When the mind of an entity perceives a threat or has unmet needs, it experiences energetic instability.

In other words, without fear (energetic instability) and evolutionary intelligence, evolution would never have happened. Energetic instability triggers evolutionary intelligence, which automatically seeks ways for an entity to regain energetic stability by neutralizing or overcoming the threat that is causing the mind of the entity to experience instability. This is true for atoms, molecules, cells, and all forms of living organisms. It is also true for *Homo sapiens* and all its group structures, including communities, organizations, and nations. If an entity cannot regain internal stability and external equilibrium, it will eventually break down into its component parts and perish.

DNA as coding

Scientists have discovered in recent years that DNA is just like computer coding. Your computer is full of coding instructions—every software program has its own set of code. However, your software doesn't work until you switch on your computer. You need an external energy source to make your software coding work. Similarly, DNA needs an external energy source to activate its coding. It needs an outside force—a life force trigger. Without an external life force trigger and energy source, DNA is inactive, just like the coding in your computer.

DNA is just coding. It needs a life force trigger to switch it on.

The thing about coding is that someone has to write it. The same is true of DNA coding. It has to be written before it can be used. This raises the question, "Who or what wrote the DNA coding that keeps organisms alive?" I believe the answer to this question is, "the evolutionary intelligence of the psychological energy force that brings conscious awareness to organisms and activates their DNA." In other words, evolutionary intelligence accounts for the psychological development of species. Furthermore, evolutionary intelligence is a property of consciousness. Wherever there is consciousness, there is evolutionary intelligence.

What I am proposing is that the soul's will to be present in 3-D awareness is the life force that triggers the DNA's coding for growth, which, in turn, triggers the DNA's coding for maintenance—the coding that allows an entity to stay alive. Once an entity reaches physical maturity, the growth coding triggers the coding for procreation—both the "physical" mechanism (puberty) and the desire for procreation (sexual pleasure).

If at any time an organism experiences a threat to staying alive that previous generations have experienced, the memory of the successful response to that threat, which is contained in the DNA's maintenance coding, is triggered, and the entity is able to take actions that successfully counter the threat. If, on the other hand, an entity experiences a threat that the species has not experienced before, then evolutionary intelligence coding is triggered.

The objective of evolutionary intelligence is to overcome the threat by creating a new set of maintenance coding. If this new coding is successful, it is integrated into the DNA's memory bank and passed down to future generations.

We can identify three sets of DNA coding that are activated by the life trigger: *growth coding, maintenance coding, and procreation coding.* The growth coding contains instructions and sequencing triggers that move an entity from an embryo to a physically mature adult. On reaching maturity, the growth coding triggers procreation coding.

Maintenance coding consists of instructions that previous generations of a species have learned about how to stay alive. Maintenance coding is triggered if the entity can give meaning to a situation that it is experiencing— if the entity can recognize the (energetic) pattern of input information (arising from the threat) and find a match in its DNA memory bank that enables it to respond to the threat in a manner that allows it to stay alive—stay present in 3-D material reality.

If an entity cannot give meaning to a situation when its survival appears to be threatened, the mind of the entity experiences energetic instability (fear), and evolutionary intelligence coding is triggered. If the evolutionary intelligence is unable to find a way to adapt or overcome the threat, then the entity will die—cease to exist in 3-D material awareness. If evolutionary intelligence is able to find a way to adapt or overcome the threat, new maintenance coding is generated and integrated into the memory bank of the DNA of that entity enabling the entity to become more resilient and therefore more likely to pass on their superior intelligence about staying alive to their progeny, resulting in natural selection.

What interests me, as far as our present inquiry is concerned, is the following question: "How does evolutionary intelligence work?" What are the algorithms that evolutionary intelligence uses to find successful solutions to staying alive when an entity encounters a situation that it, or previous generations, has never experienced before? If we can discover these algorithms, we will be able to understand the factors that have been driving the evolution of human consciousness and the whole of evolution from the Big Bang to the present day. This is the topic I propose to explore in the next chapter.

3

THE UNIVERSAL STAGES
OF EVOLUTION

IN THE LAST CHAPTER, I began to outline a theory of evolution that provides a new perspective on Neo-Darwinism and Intelligent Design—a psychological perspective that includes elements of both theories. Central to this theory is the idea of Evolutionary Intelligence—a set of problem-solving algorithms that gives purpose and direction to evolution.

As I have already stated, the purpose of evolutionary intelligence is to promote the energetic stability of entities that exist in our material world. Whenever external changes occur that threaten an entity's existence, evolutionary intelligence steps in to find ways for the entity to adapt so that it can return to energetic stability. The result of progressive adaptations overtime has led to the evolution of conscious awareness.

Evolution began immediately after the Big Bang by energy coalescing into elementary particles that were able to maintain their internal stability and external equilibrium. They became viable independent entities in their framework of existence. As their "life" conditions became more threatening, these particles bonded together to form more stable entities such as electrons, protons, and neutrons which then, as "life" conditions became even more threatening, "cooperated" with one another to form atoms. From this stable atomic (energetic) platform, life on Earth began— atoms learned how to become viable and independent. As "life" conditions continued to change, atoms learned how to bond together to form molecules, and molecules learned how to "cooperate" with one another to form complex molecules that evolved into cells.

When the "life" conditions of viable independent cells became too threatening, they bonded with other cells to form organisms. Once more,

24

as life conditions became more threatening, viable independent organisms learned how to cooperate with one another to form creatures.

One of these creatures—*Homo sapiens*—is now learning how to become viable and independent (live in internal stability and external equilibrium in its physical and cultural and frameworks of existence); bond with other members of the species to form clans, tribes, states; and nations; and nations are now learning to cooperate with one another to form higher order regional entities such as the European Union, the United States, and global entities such as the United Nations.

What is remarkable about this progression is that there was only one entity at every plane of being that was able to form a sufficiently stable energetic platform from which the next plane of being could develop: *there was only one entity at each plane of being that had the ability to cooperate.*

At the atomic plane, only the *carbon atom was able to cooperate to create higher order entities.* At the cellular plane, only the *eukaryotic cell was able to cooperate to create higher order entities.* At the plane of creatures, only *Homo sapiens was able to cooperate to create higher order entities.*

At each plane of being, there were other entities that were able to become viable and independent, and some were able to bond to form group structures, but none of these group structures were able to cooperate to form a higher order entity. This overall schema of evolution can be seen in table 3.1.

Table 3.1: Planes of being and sub-stages of evolution

Planes of being	Sub-stages of evolution	Sub-planes of being
Creatures (*Homo sapiens*)	Cooperating	Regional and global groupings
	Bonding	Clans, tribes, states, nations
	Viable and independent	Humans
Cellular (Eukaryotic cell)	Cooperating	Creatures
	Bonding	Organisms
	Viable and independent	Cells

Planes of being	Sub-stages of evolution	Sub-planes of being
Atomic (Carbon atom)	Cooperating	Complex molecules
	Bonding	Molecules
	Viable and independent	Atoms
Energetic	Cooperating	Protons and neutrons
	Bonding	Elementary particles
	Viable and independent	Energy

For each plane of being to evolve, it had to have a stable energetic platform to build on. Thus, the stability of a regional or global grouping of nations is dependent on the nations that are part of the grouping being internally stabile and in external equilibrium. If the nations forming part of a regional or global grouping are not internally stable and in external equilibrium, the regional and global grouping will not be stable. Similarly, for a nation to remain stable, the various communities that make up the nation must be internally stabile and in external equilibrium; and for a community to be stable, the families and individuals that make up the community must be internally stable and in external equilibrium. For individuals to be stable, their bodies and minds must be internally stable and in external equilibrium. For the body of an individual to be stable, the organs that make up the body must be internally stable and in external equilibrium.

For the mind of the individual to be stable, the ego and the soul of the individual must be in alignment. If the ego is unable to get its needs met, it will not be stable; it will experience fear and/or anger and will no longer be in alignment with the soul.

For the organs that make up a body to be stable, the cells that make up the organs must be internally stable and in external equilibrium. For the cells to be stable, the molecules that make up the cells must be internally stable and in external equilibrium. For the molecules to be stable, the atoms that make up the molecules must be internally stable and in external equilibrium. For the atoms to be stable, the protons, neutrons, and electrons that make up the atom must be internally stable an in external equilibrium. For the protons, neutrons, and electrons to be stable, the electronic particles that make up the protons, neutrons, and electrons must be internally stable and in external equilibrium. For the electronic particles to be stable, the energy fields of the particles must be internally stable and in external equilibrium.

Without energetic stability at the level of electronic particles, the world we live in would not exist, and conscious awareness would never have evolved. In other words, energetic stability is the foundation on which our material world is built.

If there is energetic instability at any level of being (physical or psychological), then the stability of every level above that level could be compromised. For example, if a cell in a body becomes cancerous, then the life of the body could be compromised. Similarly, if a nation in a regional group structure is unstable, then the stability of the regional group structure could be compromised.

The attributes that make evolution possible

The next important question is: "What are the attributes of these three extraordinary entities—the carbon atom, the eukaryotic cell, and *Homo sapiens*—that allowed each of them to become a stable platform for the next stage of evolution?"

The answer to this question is in three parts:

a. the ability of the entity to maintain internal stability and external equilibrium in a wide range of frameworks of existence—individual resilience;
b. the ability of the entity to bond with similar entities to create group structures—group resilience; and
c. the ability of group structures to cooperate with dissimilar group structures to form a higher order entity—collective resilience.

These are the three attributes made evolution possible and allowed consciousness to expand. Let us look at each of these entities in more detail.

The carbon atom

The *carbon atom* is one of the most stable of all elements because it has four electrons available for covalent bonding. Covalent bonding is the most resilient form of chemical bonding. It involves the *sharing* of electrons (resources) between pairs of atoms. Because of the stability afforded by this type of structural bonding, carbon was able to form durable complex molecules with many different elements.

For this reason, carbon is the second most abundant element in the human body after oxygen and the fourth most abundant element in the universe after hydrogen, helium, and oxygen. There are more compounds of carbon than all the other elements put together. Carbon atoms form the chemical basis of all forms of life known to man.

The eukaryotic cell

The *eukaryotic cell* differs from its evolutionary predecessor, the prokaryotic cell, not just because it is larger but also because of its internal structure and its ability to form communities of shared awareness. Unlike its predecessor, the prokaryotic cell, which has its "organelles" located in the cell membrane, the eukaryotic cell has its "organelles" (each organelle being a specialized prokaryotic cell) in the interior of the cell. This enables the cell membrane of the eukaryotic cell to develop more sophisticated communication systems than the prokaryotic cell. Consequently, the eukaryotic cell can bond and cooperate with other eukaryotic cells to build specialized physiological structures such as muscles, bones, and organs. Because of these attributes, eukaryotic cells are the basis of all life.

Homo sapiens

Although the jury is still out, it looks like the species *Homo sapiens* will be the fourth link in the chain of evolution, because it has a greater propensity for bonding and cooperation than any other creature.

Not only do humans have a sophisticated communication system (language), they are also able to organize themselves into communities of shared identity. We began as a species operating as family-centred clans—groups of twenty to thirty people who survived by hunting and gathering. We then formed tribes—groups of several thousand people with a shared ethnicity who survived by farming and animal husbandry. When tribes expanded into city-states, they were dominated by a single ethnic group.

However, other ethnic minorities were integrated, usually as slaves or servants.

Eventually, city-states coalesced into nations that were dominated by a single ethnic group and a single religion. Over time, ethnic and religious minorities became increasingly tolerated in nations. This is where we stand now. The most advanced nations in the world, in conscious terms, are integrating dissimilar others (ethnic and religious minorities) into their cultures. The least advanced nations still operate as human monocultures and find it very difficult to accept and integrate dissimilar others—"foreigners."

The glue that keeps communities of shared identity in a state of internal cohesion are shared beliefs, shared values, and a shared worldview. Shared beliefs and values engender trust, which is a fundamental component of social cohesion (social capital). In *Trust: The Social Virtues and the Creation of Prosperity*, Francis Fukuyama states, "One of the most important lessons we can learn from economic life is that a nation's well-being, as well as its ability to compete, is conditioned by a single, pervasive cultural characteristic: the level of trust inherent in a society."[25] As far as most nations in the world are concerned, this is a work in progress.

The most cohesive nations—the ones with the highest level of social capital—are Australia, New Zealand, Iceland, Denmark, Norway, Canada, Malta, Ireland, the US, and Indonesia. The nations with the lowest level of social capital are Georgia, Afghanistan, Chad, Mauritius, Morocco, Benin, Togo, Yemen, Central African Republic, and Burundi.[26]

The stages of evolution

We can see in table 3.1 that each plane of being can be divided into three sub-planes differentiated by scale and complexity—the plane of being of individual entities; the plane of being of the group structures that are formed when similar individual entities bond together in a shared identity; and the plane of being of similar or dissimilar groups that cooperate with one another to form a higher order identity. The higher order group structures based on the carbon

---

[25] Francis Fukuyama, *Trust: The Social Virtues and the Creation of Prosperity* (Free Press Paperbacks: New York), 1995, p. 7.
[26] The Legatum Prosperity Index. https://www.prosperity.com/rankings

atom and the eukaryotic cell were sufficiently stable to provide a foundation for the evolution of the plane of creatures and *Homo sapiens.*

We see displayed in this evolutionary progression, both in the overall schema and at each plane of being, the characteristics of increasingly higher levels of consciousness—an expansion of identity, an expansion of awareness, and a commensurate increase in the intelligence and complexity of the functioning of the mind, such that each entity could cope with the increased complexity of its existence as a member of a group or a higher order group structure. Let us now explore the characteristics of the universal stages of evolution in more detail.

Stage 1: Becoming viable and independent

To survive—stay present in our material world—the basic entities at each plane of being (elementary particles, carbon atoms, eukaryotic cells, and *Homo sapiens*) had to learn how to become viable and independent by being able to maintain their internal stability and external equilibrium in their framework of existence. When the framework conditions of an entity change, causing it to struggle to maintain its energetic stability—when it comes under threat—the first algorithm of evolutionary intelligence kicks in: the entity tries to find ways to increase its strength and resilience. If an entity cannot become viable and independent in its new framework conditions, it will eventually perish (cease to exist and disintegrate into its component parts at a lower plane of being). Thus, we can state that the fundamental principle behind the first strategy of the algorithm of evolutionary intelligence is self-interest—*caring for your own needs.*

Stage 2: Bonding to form a group structure

When the framework conditions of a viable independent entity are such that it can no longer maintain its internal stability and external equilibrium, the second strategy of the algorithm of evolutionary intelligence kicks in: the entity attempts to bond with other similar viable independent entities to share resources and form a more resilient group structure. The bonding may be temporary or permanent. If it is permanent, the individual entities will adopt the identity of the group structure.

For bonding to take place, there must be a shared purpose (usually survival), and for the bonding to become permanent there must be a set of shared beliefs and values—a shared understanding of how individual entities interact with one another to maintain the integrity (stability) of the group structure: the common good must become a higher priority than self-interest. To maintain the internal cohesion of the group structure, there must be a fair and proportional sharing of resources. If the sharing of resources is not proportional or fair, then the stability of the group structure will be compromised.

Individual members of a group structure that fail to put the needs of the group structure ahead of their own needs threaten the survival of the group structure and may potentially threaten their own survival and the potential survival of every entity that is part of the group structure. In other words, when individual entities in a group structure focus on their own "self-interest" rather than the good of the whole—when they fail to shift to a higher level of identity/awareness—the viability of the group structure is compromised. When entities bond together in a group structure, they develop a capacity for unified decision-making, which increases the group structure's ability to survive and thrive.

Thus, we can say that the fundamental principle behind the strategy of the second algorithm of evolution intelligence is *caring for your needs by caring for the needs of others with whom you share a common identity*—that is, caring for the common good. When you care for others, you can usually count on others to reciprocate by caring for you.

Entities that cannot move from self-interest to the common good are unable to maintain their stability in an increasingly threatening environment. In other words, self-interest is not a sustainable strategy in the long run, particularly in a volatile, uncertain, complex, and ambiguous framework of existence.

Stage 3: Cooperating to form a higher order entity

When the framework conditions of a group structure become overwhelming—when a group structure struggles to maintain its energetic stability (stay viable and independent)—the third strategy of the algorithm of evolutionary intelligence kicks in: group structures attempt to cooperate with other similar or dissimilar group structures to share resources, thereby forming a more resilient higher order group structure. The cooperation may be temporary or permanent, depending on the size and frequency of the occurrence of the threat.

For cooperation to take place, there must be a shared purpose (survival—physical or financial), and for the cooperation to become permanent, there must be a shared set of "values"—a shared understanding of how the group structures interact with one another to maintain the integrity (stability) of the higher order group structure.

The universal good must become a higher priority than the common good. Additionally, there must be a fair/proportionate sharing of resources between the groups that are members of the higher order group structure. If the sharing of resources is not fair or proportionate, then the stability of the higher order group structure will be compromised. When group structures do cooperate, they eventually develop a capacity for unified decision-making, which increases the higher order group structure's ability to survive and thrive.

Group structures that fail to put the needs of the higher order group structure ahead of their own needs threaten the survival of the higher order group structure and may potentially threaten their own survival. In other words, when individual group structures in a higher order group structure focus on their own "self-interest" rather than the universal good—when the members of a group structure fail to shift to a higher level of identity/awareness—the viability of the higher order group structure may become compromised.

Thus, we can say that the fundamental principle behind the third strategy of the algorithm of evolutionary intelligence is *caring for your group's needs by caring for the needs of other groups with whom you share a common identity*—that is, caring for the universal good. When your group cares for the well-being of another group, the other group reciprocates by caring for the well-being of your group. In reality, you are still caring for your own needs—cooperating becomes a way of caring for your own self-interest.

Overview

The first algorithm—becoming viable and independent—is about caring for the needs of "self," which is equivalent to "self-interest." The second algorithm—bonding to form a group structure—is about caring for the needs of "self" by caring for the needs of other "selves" who are similar and share the same sense identity. The third algorithm—cooperating to form a higher order group structure—is about caring for the needs of "self" and the needs of the group structure you primarily identify with, whilst also caring for the needs of other similar or dissimilar group structures which share the same higher order sense of identity.

In other words, evolution progresses in each plane of being by entities becoming individually more resilient, collectively more resilient, and then universally more resilient. It is not enough to learn how to be tough and strong, you must also learn how to bond with others in a group structure, and the group structure you belong to must learn how to cooperate with other group structures.

The concept of the *survival of the fittest* only correlates with the first strategy of the algorithm of evolutionary intelligence—becoming viable and independent.[27] As life conditions become more threatening, this strategy on its own does not perform as well as the second strategy—bonding to form a more resilient group structure. This finding is backed up by the latest scientific research. Using game theory, two evolutionary biology researchers found that "evolution will punish you if you are selfish and mean. For a short time and against a specific set of opponents, some selfish organisms may come out ahead. But selfishness isn't evolutionarily sustainable."[28]

Multilevel selection theory

My theory of evolutionary intelligence aligns with multilevel selection theory. In the multilevel selection theory, natural selection acts at the level of the group rather than the level of the individual. This theory was first proposed by the evolutionary biologists David Sloan Wilson and Elliott Sober. They do not posit evolution on the level of the species but at the level of groups within species. In other words, groups that bond and cooperate better survive and thrive better than groups that do not.

---

[27] "Survival of the fittest" is a phrase that originated from Darwinian evolutionary theory as a way of describing the mechanism of natural selection. The concept of "fitness" refers to reproductive success.

[28] Christoph Adami and Arend Hintze, "Evolutionary Instability of Zero-Determinant Strategies Demonstrates that Winning Is Not Everything," *Nature Communications* 4:2193, doi: 10.1038/ncomms3193 (2013).

PART 2

UNDERSTANDING PERSONAL EVOLUTION

4

HUMAN DEVELOPMENT

NOW THAT WE HAVE A clear understanding of what evolutionary intelligence is, and how it operates, we can begin to explore how the algorithms of evolutionary intelligence influence the personal development of *Homo sapiens*.

Overview

There are many models of human development, each of which describes the process of psychological growth in slightly different ways.[29], [30] The Barrett Seven Stages of Psychological Development Model is similar to most other developmental models but differs in one important way. It looks at the human psyche[31] through the lens of *ego-soul dynamics*—the growth and development of the ego, the alignment of the ego with the soul, and the activation of the soul consciousness. Almost all other models of human development ignore the soul. The main reason for this is that these models have been developed in academia, where the soul is persona non grata—a person not appreciated.

---

[29] For a list of development models, see Ken Wilber, *Integral Psychology: Consciousness, Spirit, Psychology, Therapy* (Boston: Shambhala Publications), 2000; and Dr. Alan Watkins, *Coherence: The Secret Science of Brilliant Leadership* (London: Kogan Page), 2014.

[30] You can also find a discussion of six models of maturation in George E. Vaillant, *Triumphs of Experience* (Boston: First Harvard University Press), 2012, pp. 114–189.

[31] See the appendix for an explanation of this term.

My book, *A New Psychology of Human Well-Being: An Exploration of the Impact of Ego-Soul Dynamics on Mental and Physical Health* provides a detailed explanation of the seven stages of psychological development.[32] Figure 4.1 shows the relationship between the seven stages of personal psychological development and the three levels of evolutionary intelligence (EI).

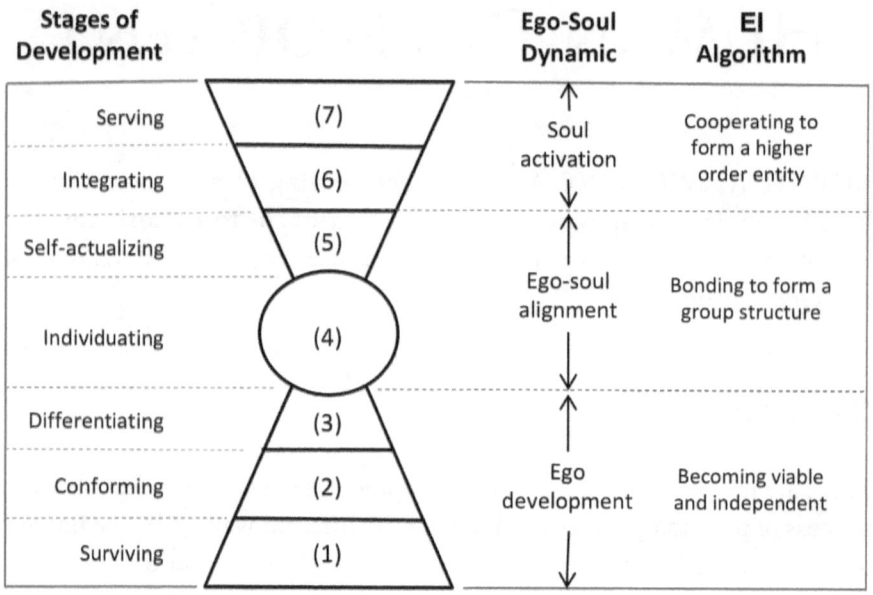

Figure 4.1: The seven stages of personal development

Ego development—becoming viable and independent

Between the moment we are born and the time we reach physical maturity, around twenty to twenty-five years of age, we pass through three stages of psychological development: surviving, conforming, and differentiating. These stages are primarily driven by our biological development—the development of our triune brain[33] and the socialization needs thrust upon us by our parents and the culture of the community in which we live. By

---

[32] Richard Barrett, *A New Psychology of Human Well-Being: An Exploration of the Impact of Ego-Soul Dynamics on Mental and Physical Health* (Fulfilling Books: London), 2016.

[33] See the appendix for an explanation of this term.

the end of the third stage of development, we are either aligned with the culture and worldview of our parents and the community/nation in which we live, or we have rebelled in some way because the parental and cultural environment we were raised in failed us—it did not allow us to get our deficiency needs met.

Surviving

From the moment we are conceived, up until the third month of gestation, our embryo lives in a state of soul awareness. Our soul has decided to experience 3-D material awareness—it has incarnated into a human body but is still living in a world of oneness, connection, and love—the world of 4-D energetic awareness.

Around the end of third month of gestation, the reptilian mind/brain (body mind), which has been forming in the background, takes over from the soul mind as the dominant mind and begins to regulate the functioning of the body. At this moment, the soul mind becomes the subconscious of the body mind. The job of the reptilian mind/brain is to keep the body alive by regulating the body's homeostatic functions, thereby maintaining internal stability and external equilibrium.

The driving force behind the reptilian mind/brain's will to stay alive is the soul's will to be present in material awareness. If the reptilian mind/brain is unable to maintain the internal stability of the body, then the body will no longer be able to act as a vehicle for soul consciousness, and the body will die.

The reptilian mind/brain stays dominant for about eighteen to twenty-four months after birth. This period—from conception to around the age of two—is what I refer to as the surviving stage of development.

By the time the infant reaches twenty-four months, it is fully operational, and it has developed a sense of its own agency. During this period the feeling of connectedness that was present during the first few months of life gives way to the feeling of separation.

For the soul, the feeling of separation is painful. It is the antithesis of its energetic experience of love. To protect itself from the pain of separation, it creates the psychic entity we call the ego. The purpose of the ego is twofold: to protect the soul from the pain of separation and to support the soul in being present in material awareness.

Conforming

The limbic mind/brain (the emotional mind), which has been growing and developing in the background, begins to take over from the reptilian mind/brain as the dominant mind/brain around eighteen to twenty-four months, at the same time as the ego begins to form. At this point, the reptilian mind/brain (body mind) becomes the subconscious of the emotional mind, and the soul mind becomes the unconscious of the emotional mind. During the conforming stage (two to seven years of age), when the emotional mind/brain is dominant but still growing and developing, the ego is focused on getting its survival and safety needs met within the parental and familial framework of the child's existence; it is looking for physical nurturance, protection, and safety—love and connection, which translates into a sense of belonging. Along with the sense of belonging, the child begins to form a sense of identity.

If the experiences of the young child are too painful for the ego to bear, because of physical or emotional abuse, the ego mind creates an alter ego to protect itself. The presence of an alter ego affects the functioning of the whole body. For example, when an alter ego is dominant, the body may show signs of diabetes or even blindness. When the normal ego returns, these symptoms go away.

Differentiating

Around the age of seven or eight, the neocortex mind/brain (the rational mind), which has been developing in the background, becomes dominant and goes on growing and developing until we reach our early twenties. This is known as the differentiating stage of development. During this stage of development, the ego is focused on getting its security needs within its peer group and its cultural environment—it is looking for respect and recognition. It wants to experience a feeling of competence, which gives it a sense of self-confidence. During this period of development, the emotional mind becomes the subconscious of the rational mind; the body mind becomes the unconscious, and the soul mind becomes the super unconscious. The influence of the ego mind on our decision-making is at its highest point, and the influence of the soul mind is at its lowest point.

What we are attempting to do during the surviving, conforming, and differentiating stages of psychological development is to establish ourselves as viable independent entities in the social and cultural framework of our existence by learning how to gratify our deficiency needs—survive, keep safe, and feel secure.

Overview

If you grew up in challenging physical, social, and cultural conditions, and you developed fear-based beliefs about not having enough to survive, not being loved enough to feel safe, and/or and not being recognized enough to feel secure, you will find it difficult to let go of these fears. Consequently, you will stay focused on the first three stages of psychological development throughout your life and not be able to align your ego motivations with your soul motivations by moving through the individuating stage of development. You will remain emotionally unstable.

Under such conditions, your evolutionary intelligence will always revert to the first algorithm of evolutionary intelligence to resolve the issues you face in your life. When you encounter a threatening situation, you have never experienced before, your evolutionary intelligence will resort to the algorithm of self-interest—attempting to become more viable and independent. Your motivations will always be selfish. You will always care about your needs more than you care about the needs of others. At the limit, you may become a sociopath or psychopath.

On the other hand, if you grew up in a safe physical environment and a loving and respectful parental, social, and cultural environment, without experiencing too many life-threatening traumatic experiences, you will not have developed too many subconscious fear-based beliefs and will find it relatively easy to master your deficiency needs and thereby become viable and independent in your framework of existence. In this case, you will not have any great difficulties moving to the individuating stage of psychological development, where you can learn to blend your ego motivations with your soul motivations.

Ego-soul alignment—bonding to form a group structure

The process of ego-soul alignment—the bonding of the ego with the soul—corresponds to the individuating and self-actualizing stages of psychological development. Unlike the first three stages of psychological development, the individuating and self-actualizing stages of psychological development are not thrust upon you by the biological and societal exigencies of growing up: these stages of psychological development are driven by the evolutionary impulse of your soul to find internal stability. You do this by releasing the ego's fears and blending your ego's motivations with your soul's motivations.

When you reach the individuating and self-actualizing stages of psychological development, your evolutionary intelligence will resort to the second algorithm of evolutionary intelligence—bonding to form a group structure. Your focus will be on the common good. You do this both internally and externally.

Internally, you first learn how to bond your ego motivations with your soul motivations (individuating) and then you learn to explore your innate gifts and talents (self-actualizing).

Externally, you first learn how to bond with other people in your workplace to become part of a working team; and you learn to how to bond with other people in your living environment to become part of a community. Then you learn how to express your unique gifts and talents to find meaning and purpose in your life.

Most adults never get to the individuating stage of psychological development. They either find it difficult to master their survival, safety, or security needs; or, because of poor parenting, they were unable to satisfy their deficiency needs. Their fear-based beliefs about surviving, keeping safe, and feeling secure dominate their minds and keep them anchored at the lower levels of consciousness. Alternatively, they may live in authoritarian regimes where freedom of expression—being who you really are—is actively discouraged.

What you are attempting to do at the individuating stage of development is dis-embed yourself from the parental and cultural dependencies of your upbringing by finding freedom and autonomy—to become fully accountable and responsible for your life. Only when you have found this freedom and autonomy can you self-actualize—truly discover who you are and begin the process of becoming one with your soul.

This process can involve a shift in your sense of identity. If you do not feel aligned with the values and beliefs of your parents or your community, you must find a new identity. This can be a difficult time in your life. In the time between letting go of your old identity, and finding and embracing a new identity, you can become vulnerable and insecure. Finding an answer to the question "Who am I?" becomes an important priority. This may lead you on a spiritual quest.

Assuming you successfully complete the individuating stage of development, when you get to the self-actualizing stage of development you will be seeking to learn how to fully express who you are so you can find meaning and purpose in your life. If you have any fears left over from the surviving stage of development—fears regarding your ability to make ends meet—they tend to show up at this stage of development. Why? Because following your passion—embracing your gifts and talents—may not be financially rewarding enough to allay your fears of survival.

Soul activation—cooperating to form a higher order entity

Once you have mastered your deficiency needs, have individuated (found your true sense of identity), and self-actualized (learned how to express your innate gifts and talents) you are ready to proceed to the integrating and serving stages of psychological development.

At the integrating stage of development, you learn how to connect with others in unconditional loving relationships, so you can actualize your gifts and talents and thereby make a difference in your world. If you have any leftover fears from the conforming stage of development—fears regarding your ability to form loving relationships—they will show up at this stage of development. A lack of emotional and social intelligence is a severe handicap for integrating.

At the serving stage of development, you learn how to contribute to the well-being of humanity and the planet. If you have any leftover fears from the differentiating stage of development—fears regarding your self-confidence and competence—they will show up at this stage of development. A lack of trust in your abilities and competence is a severe handicap for serving others.

When you reach the integrating and serving stages of psychological development, you will resort to the third algorithm of evolutionary intelligence—cooperating with other similar or dissimilar groups to form a higher order group structure. You begin to focus on the universal good. You do this internally and externally.

Internally, you learn how to become one with your soul. First you learn how to trust your soul (integrating) and then you learn how to become one with your soul (serving). Eventually, you learn how to become the servant of your soul.

Externally, you learn how to serve the universal good. First, you learn how to support your group in cooperating with other groups to form a higher order group structure; and second, you learn how to leverage your unique gifts through the higher order group structure, so you can increase your impact on the world. People working in teams learn how to cooperate with other teams to form a successful organization. People working together in communities learn how to cooperate with other communities to form a successful nation. People working together in nations learn how to cooperate with other nations to form successful regional or global organizations.

The seven stages of psychological development

With this brief overview of the stages of individual psychological development and their relationship to evolutionary intelligence, let us now explore in detail each stage of psychological development.[34] I apologize in advance that in this section you will find some of the same information I presented at the start of this chapter.

Surviving

For the first three months of life, from the moment of conception to the formation of the reptilian mind/brain, the soul mind is the dominant (conscious) interface with the embryo's external world—the mother's womb. The species mind, which is the subconscious of the soul mind, guides the development of the embryo into a fetus and creates a functioning body mind (reptilian mind/brain) by around the first trimester of gestation.

When the body mind becomes operational, it becomes the new conscious interface with the external world, and the soul mind becomes

---

[34] Throughout most of the remainder of the book, I will be shortening the term "seven stages of psychological development" to the "seven stages of development."

the subconscious of the body mind. The species mind (encapsulated in the coding of DNA) then becomes the unconscious of the body mind, guiding the development of the body through to maturity.

When the reptilian mind/brain is dominant in utero and during the first two years of your life, any difficulties experienced by the fetus or baby may influence the future functioning of the body. This leads to the phenomena known as epigenetics,[35] where the expectant mother's experiences alter the expression of the DNA coding (not the actual coding) of her unborn child.

For example, during World War II, in the Dutch famine of 1944, thousands of mothers experienced harsh deprivations that affected their unborn children. Not only did these children grow up to be smaller than average, later on, the children of these children were also smaller than average—suggesting a DNA link. Over their lifetimes, the children who lived through the period of famine in utero experienced far-above-average rates of obesity, type 2 diabetes, cardiovascular problems, and other diseases related to an unhealthy body weight.

The primary focus of the body mind is staying alive. It does this by regulating the body's internal stability. Because of our species programming (DNA), the body mind instinctively knows how to manage the body's homeostatic functioning, and once the baby is born, it knows how to suckle and knows how to cry if it feels discomforting sensations.

The embryo, the fetus, or the baby is completely dependent on the mother for its sense of well-being. The experiences that the fetus or baby have while the body mind is learning to be in 3-D material reality lead to the formation of the young child's subconscious beliefs—the body mind's autobiographical memory imprints (beliefs) that it uses to make meaning of situations that involve survival.

Although the body mind knows how to react to any internal instability, such as hunger, thirst, being too hot or too cold, it doesn't know how to alleviate these sensations. If the baby's reactions (grimacing, crying, etc.) to these discomforting sensations result in getting its needs met, it feels loved. If, on the other hand, its reactions go unnoticed or are ignored, it becomes increasingly distressed; it becomes fearful and experiences a sense of disconnection—a feeling of separation.

---

[35] See the appendix for an explanation of this term.

Gradually, the baby begins to link its reactions to its sensations, to the response it gets. It realizes that through its reactions, it is able, or as the case may be, not able, to control its sense of well-being. If its needs are "magically" met, it feels in control of its world. If its needs are not met, it does not feel in control of its world—it feels fearful.

The problem the fetus or the newly born baby has is that it still believes it is living in an energetic field of connectedness and love, because it has not yet learned about separation. The baby gradually learns through the experience of uncomfortable sensations that it is no longer living in that world, and it begins to fear separation. For the soul, the feeling of separation equates to a lack of love. Attachment theory[36] suggests that all babies begin to experience this sense of separation somewhere between six and eighteen months.

At this point, usually around eighteen months to two years, the "pain" associated with the feeling of separation becomes too much for the soul to bear. It filters out this pain by creating the psychic entity we call the ego. The ego's role is to buffer the soul from the world of separation and enable you (the soul) to become a viable independent entity in the framework of your existence.

If the mother or caregivers of the baby are not vigilant, or if the baby is abused, left alone for long periods of time, or abandoned, the baby will form subconscious beliefs that the world it lives in is an unsafe place and that it is not loved. After that, throughout his or her life, this person will seek to control what is happening in their environment to make sure they get their needs get met. As an adult, such a person will be cautious and vigilant and tend to micromanage or control whatever is going on around them that might affect their well-being.

If, on the other hand, the mother or caregivers of the baby are attentive to its needs and are watchful and responsive to signs of distress, then the baby will grow up with the feeling of being loved and that the world it lives in is a safe place. The feeling of control the baby gets when its needs are quickly met is an essential prerequisite for mastering the self-actualization stage of development later in life. If you don't feel in control, you will not be prepared to take the survival risks that the journey into self-actualization may entail.

At the survival stage of development, love is experienced through the satisfaction of our physiological needs. This is when the body mind

---

[36] Ibid.

experiences stability. The body mind experiences instability—a lack of love—when it feels abandoned and uncared for.

Conforming

Towards the end of the surviving stage of development, the infant becomes mobile and learns to communicate verbally. This is the time when the ego begins to form and the limbic mind/brain (emotional mind), which has been developing in the background, becomes the dominant mind/brain. Whereas the focus of the body mind is on staying alive, the focus of the emotional mind is on safety and protection. The body mind goes on functioning in the background as the physical interface with the world, and the emotional mind becomes the social interface with the world.

When the emotional mind becomes dominant, the body mind becomes the subconscious of the emotional mind and the soul mind is pushed further into the background. It becomes the unconscious of the emotional mind. It still has some influence over the thoughts of the child, but these are less overriding than before.

If the child's emotional mind feels unfairly treated, instead of becoming angry with its parents, the child may repress the anger it is feeling and blame itself. The child blames itself because it is afraid to show anger towards its parents: if it does show anger towards its parents, it realizes that it might be more difficult to get its safety needs met in the future. This is when your inner critic is born—the voice of judgement about not being worthy or good enough to receive the love we are seeking. Unless it is dealt with, the emotional instability thus caused does not go away. It is always there in the background, influencing the child's and the adult's subconscious decision-making.

At the beginning of conforming stage of development, the child may resort to temper tantrums to get its needs met. The young infant has not yet learned how to separate itself from its needs. Neither has it learned that the people it depends on for its survival and safety may have competing needs. If the parents give in to the child's temper tantrums, the child quickly learns that behaving "badly" is a good strategy for getting its needs met. When this happens, the parents' lives become intolerable—they become totally ruled by their children.

Alternatively, if the parents make getting the child's needs and desires met conditional on the adherence to certain rules of behaviour—if the child is coerced into behaving in specific ways—the child will learn that love is conditional and will tend to use this strategy to manipulate others into getting its needs met later in life.

For the sake of family unity, the growth of the child's ego has to be managed. There are two ways of doing this: the correct way—by gradual socialization (getting the child to recognize that other people may also have needs); and the incorrect way—by attempting to crush the child's ego through force, punishment, or making the giving of love conditional.

If the child's parents or caregivers are attentive to the child's needs, if it is raised in a caring, loving environment where it feels safe and protected, then the child will grow up with the desire and willingness to form committed relationships and conform to society's rules when it reaches adulthood. Participating in family rituals is important at the conforming stage of development, because they contribute to the child's feeling of belonging and safety.

Learning to feel safe, comfortable, and loved at the conforming stage of development is an essential prerequisite for mastering the integrating stage of development later in life. If you don't feel safe with others—if you don't trust them—you will find it difficult to connect and cooperate with others when you become an adult.

At the conforming stage of development, love is experienced through the satisfaction of our safety and protection needs. This is when the emotional mind experiences stability. The emotional mind experiences instability—a lack of love—when its love and belonging needs are not met.

Differentiating

Towards the end of the conforming stages of development, around the age of seven or eight, the rational mind (the neocortex), which has been developing in the background, gradually takes over from the emotional mind as our conscious interface with the world. The focus of the rational mind is on security—keeping us safe from harm and supporting us in finding our place in the world. The emotional mind goes on operating in the background as our social interface with the world, and the body mind goes on operating in the background as our physical (biological) interface with the world.

When the rational mind becomes dominant, the emotional mind becomes the subconscious of the rational mind, and the body mind becomes the unconscious of the rational mind. The soul mind becomes the super unconscious. Its influence is felt only faintly, if felt at all, especially if the fears of the ego are severe. The energy of love will not be able to break through the energy of fear.

Subconscious decisions made by the emotional mind can be overridden by the rational mind if the rational mind believes the reactions of the emotional mind (anger) would compromise its ability to get its security needs met. Therefore, most people tend to demote a lower order emotional need for a higher order rational need (security). If, however, you found it difficult to get your emotional needs met when you were young, your override function may be compromised. You will lash out, paying little attention to the consequence of your outburst. Our prisons are full of people with faulty override functions.

When the child becomes a teenager, it starts to explore the world outside its family environment. Whereas parental and sibling relations were of significant importance to the child's safety up to the age of seven or eight, when the child gets close to its teens, its relations with its peers and the authority figures in its life, such as teachers or religious instructors, become important for satisfying its security needs.

The teenager gets his or her security needs met, either by associating with a community, clique or gang of peers, or by staying in close contact with its parents. To get its security needs met, the teenager must find a way to become respected—to be recognized and felt seen: it must prove it is worthy of belonging to the family or group of people it identifies with.

There are three ways for teenagers to get the respect and recognition needs met:

- **By physical body image displays:** For boys, this means becoming strong or powerful; for girls, this means becoming beautiful or sexy. This is usually the route that is taken to get our recognition needs in peer gangs or cliques.
- **By displays of knowledge and learning:** For boys and girls, this means becoming a good student and being smart. This is usually

the route that is taken to get our recognition needs met from parents and authority figures.

- **By displays of status ("coolness"):** For boys and girls, this means having the latest gadgets and the most fashionable hairstyles and clothes. This is usually the route that is taken to get our recognition needs met from our peers.

Which path or mixture of paths the teenager chooses to get its respect and recognition needs met will depend to a large extent on the relationship he or she has with his or her parents. If the relationship the teenager has with his or her parents is good, they will feel recognized and appreciated; if the relationship the teenager has with his or her parents is poor, then they will turn to an authority figure or a peer group to get their recognition needs met.

What is important at this stage of development is for teenagers to get positive feedback from their parents. If they do not get positive feedback from their parents, they will seek to get it from other people. They will join a group or gang where they feel accepted, and where their gifts, skills, or talents are recognized.

If the teenager joins a gang, taking on dares can become a rite of passage for membership of the gang. This may lead young people "off the straight and narrow." They may do things they know to be wrong simply to belong to a group where they feel recognized. Teenagers who form relationships with an adult outside the home to get their recognition needs met may leave themselves open to religious radicalization or sexual grooming.

Joining a gang may create conflicts in the teenager's life at home, because they may get caught between two value systems: the value system of their parents and the value system of the group or gang to which they belong. If this situation is not handled sensitively by parents, home life will become difficult and may become intolerable. In which case, you will have a rebellious teenager on your hands.

From a parental perspective, guiding rather than controlling, allowing rather than preventing, encouraging rather than denigrating, and trusting rather than doubting, gives teenagers space to safely explore who they are and find their sense of identity in the larger world outside the family home. The most important thing is to spend time with your teenager.

Feeling physically and emotionally secure in your community— having a healthy sense of self-esteem by being respected and recognized by others—is an essential prerequisite for mastering the serving stage of development later in life. If you don't feel secure in your community during your teenage years, you will not feel confident in contributing to society later in life.

At the differentiating stage of development, love is experienced through the satisfaction of our security needs. This is when the rational mind experiences stability. The rational mind experiences instability—a lack of love—when its respect and recognition needs are not met and/or it is made to feel incompetent.

Individuating

Around your mid-twenties, you begin to feel the need for autonomy and freedom—to break the chains of dependency that keep you tied to the parental and cultural framework of your existence. You are finished with being dependent on others for the satisfaction of your deficiency needs; you are seeking independence. You want to become responsible and accountable for every aspect of your life; in particular, you will want to explore your own beliefs, and embrace and express your values.

Fundamentally, the task at the individuating stage of development is to embark on the journey that will lead to the recovery of your soul. Without fully realizing it, you will be dis-embedding yourself from your parental and cultural background and starting to align the motivations of your ego with the motivations of your soul.

For those who were fortunate enough to have been brought up by self-actualized parents, and lived in a community or culture where freedom and independence were celebrated, where higher education was easily available, where men and women were treated equally, and where they were encouraged from a young age to express their needs and think for themselves, they will find it relatively easy to move through the individuating stage of psychological development; that is, as long as they can find work that enables them to make a living. If you cannot find work that gives you financial independence, you will feel demoralized, because you will not have the autonomy and freedom you need to individuate. You will still be under the control of your parents, because they will be paying your bills.

Many find it difficult to extract themselves from the influence of their parents, even when they are financially independent. Others, such as those who live in authoritarian communities or repressive regimes, may be afraid to express themselves, because they fear being punished for breaking the rules or don't want to be locked up for speaking their truth. Thus, if you were brought up by controlling parents, if you live in an authoritarian regime, if you are discriminated against because of your gender, sexual preferences, religion, or race, and you have fears about being able to meet your deficiency needs, you are likely to have difficulties moving through the individuating stage of development. Your fears will keep you anchored in the lower levels of consciousness.

Thus, your task at the individuating stage of development is to master these fears so you are no longer dependent on others for your self-esteem, protection, and survival. If you do not overcome these fears, they will continue to show up in your life as an adult and make it extremely difficult to master the higher stages of development. Your soul needs you to let go of your parental programming and cultural conditioning, so you can fully express yourself.

Self-actualizing

If you successfully master the individuating stage of development, around the time you reach your forties, sometimes a little earlier and sometimes a little later, you will experience the soul's desire for self-expression. You will want to find a meaning and purpose to your life. You will be looking for a vocation or calling that aligns with your soul's purpose. This means uncovering your natural gifts and talents and making them available to the world.

For most people, finding their vocation or calling usually begins with a feeling of unease or boredom about their job, profession, or chosen career—with the work they thought would enable them to feel secure by providing them with a good income and prospects for advancement leading to increased wealth, authority, or power.

Uncovering your soul's purpose not only brings vitality to your life, it also sparks your creativity. You will become more intuitive and spend more time in a state of flow—being present in what you are doing and feeling committed and passionate about your work.

Mastering the self-actualizing stage of development can be challenging, especially if your vocation or calling offers less security than the job, profession, or career you trained for earlier in your life. You may feel scared or uncomfortable about embarking in a new direction that does not pay the rent, the mortgage or finance your children's education but does bring meaning and purpose to your life. This is why it so important at this stage of development to master your survival fears. Knowing you can take care of yourself gives you the confidence you need to explore your self-expression. If you are afraid that you might not be able to survive doing what you love to do, you may deny your soul expression. This will lead to mental suffering, culminating in depression. Uncovering and embracing your soul's purpose is vitally important, because it is the key to living a fulfilling life.

Some people find their vocation early in their lives; others discover it much later; some spend their whole lives searching. I often tell people who are having difficulty finding meaning and purpose not to worry. Only the ego is concerned about meaning and purpose: the soul is concerned about self-expression. Just do what you love to do—do what brings you joy. Full self-expression is vital for living in soul consciousness.

Integrating

If you were successful in traversing the individuating stage and found your soul's purpose at the self-actualizing stage, when you reach your fifties you will want to use your gifts and talents to make a difference in the world. To do this, you will need to form caring relationships with those you want to help and those you want to collaborate with to leverage your impact in the world. Connecting with others who share your passion or calling and with those who will be the beneficiaries of your gifts and talents is essential for mastering this stage of development.

To connect with and support others, you will need to tap into your emotional and social intelligence and exercise your empathy skills. You will need to feel what others are feeling if you are truly going to help them. Thus, how well you mastered the conforming stage of development will significantly influence your progress through the integrating stage of development.

Knowing you can handle your relationship needs—knowing you are lovable and can love others—gives you the confidence you need to successfully manage the integrating stage of development. In addition, you must also be able to cooperate with others by assuming a larger sense of identity and shift from operating independently to operating interdependently.

Some people get so wrapped up in their "work" at the self-actualizing stage that they are unable to make the shift to the integrating stage. They get lost in their creativity, focusing only on the expression of their gifts and talents, rather than the larger contribution they could make if they were able to connect and collaborate with others. Working with others in service to the universal good is more likely to bring a sense of fulfillment than working on your own at this stage of your life.

I tell people who are finding it difficult to make a difference in the world not to worry. Only the ego is concerned with making a difference: the soul is concerned with connecting. You must first learn how to connect in unconditional loving relationships; only then can you make a difference in the world.

Serving

The last stage of development follows naturally from the integrating stage. I call this the serving stage of development. This stage of development usually begins in the early sixties, sometimes a little earlier, sometimes a little later. The focus of this stage of development is on selfless service to the community you identify with. What you are feeling is the soul's desire for contribution.

How well you mastered the differentiating stage of development will significantly influence your progress through the serving stage of development. Having a healthy sense of self-esteem and self-confidence will enable you to make your gifts and talents available to those who need them.

It does not matter how big or small your contribution, what is important is fulfilling your soul's purpose. Alleviating suffering, caring for the disadvantaged, and building a more loving society are some of the activities you may want to explore at this stage of your life. On the other hand, your contribution may be simply caring for the life of another soul.

As you enter the serving stage of development, you will find yourself becoming more introspective and reflective—looking for ways to deepen your sense of connection to your soul and beyond your soul to the deeper levels of your being—connecting to whatever you consider to be the divine. You may become a keeper of wisdom, an elder of the community, or a person to whom younger people turn for guidance or mentoring.

As you make progress with this stage of development, you will uncover new levels of compassion in your life. You will experience feelings of well-being and fulfillment that you never experienced before. You will begin to see how connected we all are; how, by serving others, you are serving your larger self. At this level of consciousness, giving becomes the same as receiving. When you give to others, you are giving to yourself. To experience these feeling more profoundly, you will want to become the servant of your soul. Eventually, you may realize that you don't *have* a soul, you *are* your soul; you are living the life of an energetic being in three-dimensional material awareness.

I tell people who are finding it difficult to master the serving stage of development not to worry about service. Only the ego is concerned with being of service: the soul is concerned about making a contribution. You must find ways to contribute to the well-being of others. If you cannot find ways of contributing to the well-being of others, your life is basically over. Playing endless rounds of golf to lower your handicap, doesn't really do it for the soul.

5

Notes on the Stages of Psychological Development

I THINK IT WOULD BE useful at this point to provide more information about how the seven stages of psychological development operate and reemphasize the link between the algorithms of evolutionary intelligence and the stages of psychological development. There are four topics I want to cover:

- The link between the algorithms of evolutionary intelligence and stages of development.
- The ordering of the stages of development.
- The link between the stages of development and levels of consciousness.
- What happens if we fail to master a stage of development.

The link between algorithms of evolutionary intelligence and stages of development

The first algorithm—increasing your strength and resilience by becoming viable and independent—operates at the first three stages of development, during the period when the ego is dominant. If a person never succeeds in individuating, the ego will always stay dominant, and the individual will always use the first algorithm of evolutionary intelligence to find ways to respond to situations they have never experienced before when their needs are under threat. Such individuals will never find fulfillment in their lives. They may experience "success," but they will always feel internally unstable; they will never be at peace, because their minds will be eternally vigilant looking for opportunities to get their deficiency needs met.

The second algorithm—increasing your strength and resilience by bonding with similar others in a group structure—operates at the fourth and fifth stages of development, during the period when we are learning to blend our ego motivations with our soul motivations by letting go of our fears and uncovering our innate gifts and talents—finding out and embracing who we really are. Success at these stages of development fundamentally depends on bonding with your soul, either consciously or subconsciously, and bonding with others.

The third algorithm—increasing your strength and resilience and the strength and resilience of the group you identify with by forming a higher order group structure—operates at the fifth and sixth stages of development, during the period when we are learning how to activate our soul by letting go of our fears of connecting and contributing. This is the period when you fully step into soul consciousness. You are attempting to recreate your soul's natural four-dimensional energetic experience in your three-dimensional material world.

The ordering of the stages of development

The seven stages of psychological development occur in consecutive order. Each stage of development provides a foundation for the subsequent stage. You cannot jump stages, but from time to time you may experience higher *states* of consciousness, especially after you have reached the individuating stage of development.

You must establish a stable energetic platform at each stage of development to proceed to the next. For most people, it takes a lifetime— at least sixty to seventy years—to pass through the seven stages of development. That is because each stage is linked to the seasons of our lives. Most people, however, never manage to reach the individuating stage of development.

If you complete the journey through to the serving stage, you can look forward to flourishing in the latter years of your life. You will feel a deep sense of joy; you will experience good health and have a long life. The approximate age range of each stage of development, the tasks we must master, and the needs we must meet are shown in table 5.1

Table 5.1: The tasks and needs associated with the
seven stages of psychological development

Stage of development	Age range	Task	Need
Serving	60+ years	Contributing to the well-being of others.	To be of service to humanity.
Integrating	50–59 years	Connecting in unconditional loving relationships.	To make a difference in your world.
Self-actualizing	40–49 years	Discovering and expressing your gifts and talents.	To find meaning and purpose.
Individuating	25–39 years	Finding freedom to discover who you are.	To feel you can operate with autonomy.
Differentiating	8–24 years	Feeling accepted, respected, and recognized.	To feel a sense of self-worth.
Conforming	2–8 years	Feeling safe, protected, and loved.	To feel a sense of belonging.
Surviving	0–2 years	Feeling physically cared for and nurtured.	To feel in control of your life.

The first three stages of development are fixed in time. They are determined by the physical development of our brains and bodies and the mental development of our minds. Thereafter, if all goes well, it is possible to accelerate the stages of development, but it is quite rare. Many young people these days think they have reached the integrating stage of development because they want to make a difference in the world.

When someone is truly at the integrating stage of development, the desire to make a difference flows from their sense of empathy for the plight of others. Usually, this is not the case with millennials: their desire to make a difference comes either from their need for achievement or their desire for justice—to right the wrongs they see in the world. The hallmarks of the integrating stage are empathy for others and connecting in unconditional loving relationships.

No matter what age you are, to grow and develop you must address the fears and anxieties you have about being able to meet your deficiency (survival, safety, and security) needs—the unmet needs you may still have from the first three stages of development. Also, if you are at the individuating stage of development, you must learn to master your fear of leaving those you are dependent on—your parents and the community in which you were brought up—to find the freedom and autonomy you need to discover who you really are. If you were brought up by self-actualized parents, this will not be very difficult, because they will understand your need for autonomy. If you weren't, you may find it challenging. Your parents may not understand why you are distancing yourself from them; why you have a different worldview. You may be wondering why you no longer resonate with them. You may even feel guilt. Let the guilt go. Recognize they are on a different journey to yours. Stay loving and dutiful, but don't beat yourself up.

The link between the stages of development and levels of consciousness

We grow in stages, and we operate at levels of consciousness. Normally, the level of consciousness you operate from will be the same as the stage of development you have reached. However, when your mind interprets a current experience as potentially threatening to getting your survival, safety, or security needs met, through its conscious or subconscious fear-based beliefs, you will drop down to one of the first three levels of consciousness.

This does not mean you are moving to a lower stage of development. It simply means that you are moving to a level of consciousness where you are facing similar issues to those you had when you were at that stage of development. For example, if I am thirty-five years old and at the individuating stage of development and lose all my money, I will immediately descend to the surviving level of consciousness. Similarly, if I am single and move to a foreign country, I will want to make friends in my new environment. I will automatically descend to the relationship level of consciousness. In these instances, I do not go back to the surviving stage of development or the conforming stage of development, instead I return to the survival and relationship levels of consciousness. The correspondence between the seven stages of development and the Seven Levels of Consciousness are shown in figure 5.1.

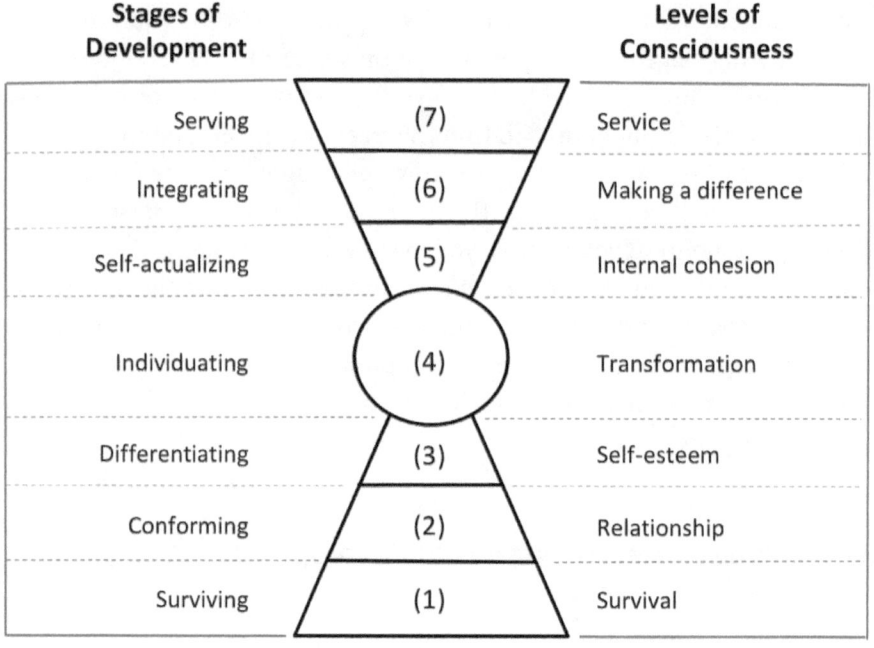

Figure 5.1: The seven stages of development
and the seven levels of consciousness

What happens if we fail to master a stage of development

The first three stages of development are intimately linked to the last three stages of development. If you are at the self-actualizing stage of development—finding and engaging in the work you are passionate about—you must be able to master your survival fears; you must feel you are in control of your life and not a victim. If you are at the integrating stage of development—connecting with others in unconditional loving relationships to make a difference—you must be able to master your relationship fears; you must feel loved enough to feel safe in loving others. If you are at the serving stage of development—contributing to the well-being of society—you must be able to master your self-esteem fears; you must feel confident enough to go out into the world and offer your gifts and talents. Thus, how well we master the first three stages of development can impact our success in mastering the last three stages of development. Table 5.2 shows the concerns that we have and the feelings we experience when we fail to master each stage of development.

Table 5.2: Concerns and feelings associated with
the mastery of each stage of development

Stage of development	Concerns	Feelings
Serving	Lack of confidence and isolation	Nobody cares about me. I feel I have nothing to offer my community and those around me.
Integrating	Lack of connection and loneliness	I feel isolated. I have nothing in common with the people around me.
Self-actualizing	Lack of purpose/ meaning in life	I feel my life is meaningless. I don't know what my gifts and talents are.
Individuating	Lack of freedom and autonomy	I feel trapped. I can't find a way to discover who I am outside my parental and cultural conditioning.
Differentiating	Lack of recognition in your community	I don't feel seen or heard. I don't belong anywhere. I am not enough.
Conforming	Lack of harmony in relationships	I feel unsafe and unprotected. I am not loved enough.
Surviving	Lack of health or income	I feel vulnerable. I do not have what I need to survive.

Repeated painful experiences of not getting your deficiency (survival, safety, and security) needs met when you are young get "hard-wired" into your mind as fear-based beliefs, because the first twenty-four years of your life is a time of rapid emergent learning[37] when the synapses in your brain are forming.

Synapses are "electrical" connections that correspond to the beliefs in your mind. Your synapses enable you to interpret your experiences by giving them meaning. The meaning you give to your experiences aligns with your strongest synaptic connections (beliefs). Consequently, if you constantly failed to get your needs met when you were young, during the period of emergent learning, you will have formed limiting beliefs (strong synaptic connections) that may haunt you for the rest of your life. You must build new synaptic connections if you want to change your beliefs and thereby change your life.

---

[37] Ibid.

PART 3

A WAY FORWARD

6

HOW ARE WE
PROGRESSING?

What is an evolutionary human?

AN EVOLUTIONARY HUMAN IS A person who is consciously learning how to master the seven stages of psychological development by using the three algorithms of evolutionary intelligence to help them make decisions.

Evolutionary humans use the first algorithm of evolutionary intelligence—becoming viable and independent—to help them make decisions during the ego stages of development—from birth to at least their mid-twenties. Most people never get past these stages of development because they are unable to satisfy their deficiency needs, or they live in autocratic regimes where freedom of expression, and therefore individuation, is suppressed by the State.

Evolutionary humans use the second algorithm of evolutionary intelligence—bonding to form a group structure—to help them make decisions while they are learning how to blend their ego motivations with their soul motivations—from their mid-twenties through their forties, during the individuating and self-actualizing stages of development. They may also use the first algorithm of evolutionary intelligence in situations that support their personal growth and development: caring for self is just as important as caring for others.

Evolutionary humans use the third algorithm of evolutionary intelligence—cooperating to form a higher order entity—to help them make decisions from their late forties onwards—during the stages of soul activation (the integrating and serving stages of development). They may also use the first algorithm in situations that support their personal growth and the second algorithm in situations that support the growth and development of the human group structures with which they primarily identify.

As evolutionary humans learn to master each stage of development and each algorithm of evolutionary intelligence, they expand their level of conscious awareness, they expand their identity, and they expand the complexity of their mind functioning.

So how are we as a species progressing in nurturing evolutionary humans?

At this point in time, there are seven nations that have developed a worldview and a culture that nurtures human evolution. Not everyone in these nations can be called an evolutionary human, but the cultural infrastructure for making this happen is in place. These nations—Finland, Norway, Denmark, Switzerland, Sweden, Iceland, and New Zealand—are among the happiest and most internally cohesive nations on the planet. Table 6.1 compares the top seven most conscious nations with the top seven happiest nations. With one exception—New Zealand—the top seven most conscious nations[38] are also the top seven happiest nations. New Zealand, which ranks number 7 in consciousness, ranks number 8 in happiness. Canada which ranks number 7 in happiness, ranks number 8 in consciousness.

Table 6.1: The most conscious and the happiest nations

Ranking	The most conscious nations (2017)[39]	The happiest nations (2018)[40]
1	Finland	Finland
2	Norway	Norway
3	Denmark	Denmark
4	Switzerland	Iceland
5	Sweden	Switzerland
6	Iceland	Netherlands
7	New Zealand	Canada

[38] In 2017, I developed a way of measuring the consciousness of nations. This is briefly explained in my book *Everything I Have Learned About Values* pp. 82–85.
[39] https://www.aahv.global/global-consciousness-indicator.html
[40] The World Happiness Report for 2018.

The majority of people living in the seven most conscious nations are using the second algorithm of evolutionary intelligence to guide their decision-making. They are putting their nation's needs—the needs of similar others—ahead of their own needs. Additionally, we also find a strong expression of the third algorithm of evolutionary intelligence in these nations. They are also caring for the needs of dissimilar others. Evidence of the use of third algorithm shows up in the official development assistance (foreign aid spending) data (see table 6.2)

Table 6.2: Net official development assistance
by country as a percentage of GNI

Nations	Percent of GNI
Sweden	1.40%
Norway	1.05%
Luxembourg	0.93%
Denmark	0.85%
Netherlands	0.71%
UK	0.56%
Finland	0.52%
Switzerland	0.52%
Germany	0.42%
Belgium	0.37%

GNI = Gross National Income

Five of top ten foreign aid donors (as a percentage of their GNI)—Sweden, Norway, Denmark, Finland, and Switzerland—are among the top seven most conscious nations. The other five top ten foreign aid donors—Luxembourg, the Netherlands, the UK, Germany, and Belgium—are among the top sixteen most conscious nations.

In addition to the top seven most conscious nations, there are another nine nations that are making good progress in creating the cultural conditions that nurture the evolutionary human. These are Canada, Australia, Luxembourg, Ireland, the Netherlands, Germany, Austria, the UK, and Belgium. All these nations are among the top sixteen foreign aid donors, suggesting that they are also using the third algorithm of evolutionary intelligence in making decisions. Of these nine nations, Canada and Australia are the most advanced in nurturing the evolutionary human. The least advanced of these nine nations are the UK and Belgium.

In the last few years, there has been a backsliding in conscious in the UK from the use of third algorithm to the second algorithm of evolutionary intelligence. This is one of the key factors that resulted in the phenomenon called BREXIT. The main BREXIT issue is immigration—more people in the UK want to bond with similar others, and fewer people want to cooperate with dissimilar others. This phenomenon is most apparent among rural communities, older segments of the population and young people who have unmet deficiency needs. We also find this backsliding in the US among similar groups of people. This has resulted in the election of one the least conscious presidents in recent American history.

7

THE NORDIC SECRET

AT THIS POINT, THE QUESTION that arises in my mind is: "How did the Nordic nations—Norway, Sweden, and Denmark, as well as Finland and Iceland—develop a worldview/national culture that nurtures the evolutionary human?" From being at the very bottom of the European economy in the 1860s, they climbed to the top by the 1930s and have remained prosperous and progressive ever since. Finland made a similar journey after 1918 in just thirty years. We can find a response to this question in *The Nordic Secret*.[41]

The authors of *The Nordic Secret* make the case that the success of the Nordic countries was due to a very specific and targeted political project. The key cultural and political figures in the 1850s recognized that there was a need to emancipate the rural population. They saw that the industrial revolution would cause the feudal structures to collapse, resulting in a mass migration to the cities.

The rural workers needed more reading and writing skills if they were to become responsible citizens. More than that, these visionary leaders realized that the masses needed a sense of collective identity at the level of the nation. "They needed to develop a sense of responsibility towards self and society; they needed moral, emotional and cognitive development. They needed what is called ego-development in modern psychological terms. ... They wanted people to think, feel and act in different ways. ... They did not want to dictate what people should think or how they should

---

[41] Lene Rachel Andersen and Tomas Björkman, *The Nordic Secret: A European Story of Beauty and Freedom* (Stockholm: Fri Tanke Publishing), 2017.

act. They wanted a population that could 'author' their own lives and take part in the authoring of a new society."[42]

In my terminology, they wanted to create programs of education in their folk high schools that could enable the masses to individuate. Therefore instead of calling this program ego-development, I prefer to call it an individuation program. "What emerged was the understanding that people must be able to control their emotions, internalize the norms of society and take individual moral responsibility. In German, this kind of personal ego-development goes under the name *Bildung*."[43]

In my opinion, *Bildung* represents a system of education that takes people from the ego stages of development to the balancing of the ego's motivations with the soul's motivations; in other words, from the first three stages of psychological development (surviving, conforming, and differentiating) into and through the fourth stage of psychological development (individuating). This education system supports people in moving from the first algorithm of evolutionary intelligence as a way of making decisions to the second algorithm of evolutionary intelligence— from decision-making that focuses on "What's in it for me?" to decision-making that focuses on "What's best for the common good?"

The principal ingredients of the Nordic education program were as follows:

- Dedicated teachers.
- An understanding among the teachers of the concepts of psychological development.
- An understanding of the importance of cultural heritage, so people feel a sense of belonging at the national level.
- Basic reading and writing skills.
- Access to literature in people's mother tongue.
- Investment in meeting places for deep conversations and learning.
- Letting the teacher run with it.
- The teachers must also be learners.
- Poetry, aesthetics, and communal singing.

---

[42] Ibid., p.8.
[43] Ibid., p.9.

- Teaching what is immediately relevant to the participants—what empowers them politically and what improves their lives here and now.
- An understanding of autonomy and local self-organizing.
- Start with the sixteen to thirty age group.

This last point—a focus on the young—was, I believe, fundamentally important to the success of the *Bildung* program. This age group is at the differentiating and individuating stages of psychological development: they are asking questions about recognition, responsibility, and identity. They are searching for a larger sense of belonging beyond their local community.

The authors of *The Nordic Secret* believe that this approach, which has worked so successfully in Scandinavia, is replicable in other parts of the world. "Poor countries can make themselves rich in less than two generations if the population gets *Bildung*. ... There was nothing unique about Scandinavia in 1850 or Finland in 1918, quite the contrary: all four countries were poor, very religious and the rule was recently authoritarian. ... From the first successful folk high school in 1851 to when Denmark joined the economic elite took 50 years; Norway and Sweden made similar journeys and Finland did it in just 30 years."[44] One of the major success factors of the *Bildung* approach is that it helps to lift people out of poverty, thereby addressing their deficiency needs and provides a solid basis for helping them focus on their growth needs, and it does this in the framework of a free and democratic society.

The first three levels of consciousness represent Abraham Maslow's deficiency needs: survival (physical and economic), relationships (personal safety), and self-esteem (personal security). We get anxious and fearful if these needs are not met, but once they are met, we no longer pay attention to them.

The upper three levels of consciousness approximate to Abraham Maslow's growth needs: self-actualization (self-expression), integration (connecting to make a difference), and service (selfless service). When these needs are met, we want more.

Table 7.1 shows the nations that do the best job taking care of their citizens' deficiency needs and the best job of taking care of their citizens' growth needs. The top three nations at taking care of people's growth needs

---

[44] Ibid., p. 382.

are Finland, Iceland, and Denmark. The top three nations at taking care of people's basic needs are Switzerland, New Zealand, and Singapore.[45]

Table 7.1: The nations that are best at taking care
of people's deficiency and growth needs

World Ranking	Nations that take care of citizens' deficiency needs	Nations that take care of citizens' growth needs
1	Switzerland	Finland
2	New Zealand	Iceland
3	Singapore	Denmark
4	Denmark	Norway
5	Finland	Sweden
6	Norway	Australia
7	Sweden	Luxembourg
8	Netherlands	Switzerland
9	Canada	Canada
10	Australia	Ireland

It is interesting to note that Denmark, Sweden, and Finland figure on both lists; they are good at satisfying citizens' deficiency needs and growth needs. Other nations that do a good job in satisfying citizens' deficiency needs and growth needs include Switzerland, Canada, and Australia. Interestingly, Singapore, which does a good job of satisfying citizens' deficiency needs, does less of a good job at satisfying citizens' growth needs—it ranks number 24.

The Law of Jante

I believe another significant factor in the formation of the culture of Nordic nations is the Law of Jante. This is a code of conduct that has been part of the Danish, Swedish and Norwegian culture for centuries. The Law of Jante can be summarized in one sentence: "You are not to think you are

---

[45] https://www.aahv.global/global-consciousness-indicator.html

anyone special or that you are better than us." Its focus is on regulating the self-esteem level of consciousness. There are ten rules that make up the law of Jante. The ten rules state:

1. You are not to think you are anyone special.
2. You are not to think you are as good as we are.
3. You are not to think you are smarter than we are.
4. You are not to imagine you are smarter than we are.
5. You are not to think you know more than we do.
6. You are not to think you are more important than we are.
7. You are not to think you are good at anything.
8. You are not to laugh at us.
9. You are not to think anyone cares about you.
10. You are not to think you can teach us anything.

You can find these rules in a book written by Aksel Sandemose entitled *A Fugitive Crosses His Tracks* published in 1936.[46]

Sandemose was writing about the prevailing attitudes in his home town of Nykōbing Mors in Denmark. These attitudes are seen as diminishing individual effort, denigrating individual achievers, and placing emphasis on the harmony of the collective. The rules are regarded as a personal criticism of people who want to break out of their social groups and reach a higher position.

Whilst the rules that represent the law of Jante probably had a significant impact on helping the Nordic nations build cohesive national cultures in the past, by levelling out society and inhibiting the power of elites, nowadays the rules are coming under some criticism. The Law of Jante is perceived by modern social commentators as inhibiting individuation and self-actualization, which in turn leads to high rates of depression and suicide. There is little doubt, however, that the Law of Jante is alive and well in Nordic society and is likely to go on influencing Danish, Swedish and Norwegian cultures for several more generations.

The rules that make up the Law of Jante are similar to the rules that are found in tribal cultures but without the ethnocentric dimension. The rules in tribal cultures are primarily designed to afford protection through

---

[46] Aksel Sandemose, A Fugitive Crosses His Tracks (Denmark: Knopf), 1936.

belonging (level 2 consciousness), whereas the rules in the Law of Jante are primarily designed to create internal cohesion (level 5 consciousness).

I believe the principal impact of the Law of Jante in the Nordic nations has been, and still is, in curbing the excesses of the elites—reminding them of the importance of being part of a community of souls.

8

THE WAY FORWARD

FOR THE HUMAN SPECIES TO continue the evolutionary journey of consciousness, we first need to create a critical mass of people in every nation that can meet their deficiency needs; then we need to support them in satisfying their growth needs. In other words, we need to move the masses of every nation out of the ego stages of development—survival, safety, and security—so they can find the freedom they need to individuate and focus on their growth needs. Once we have given people the possibility of individuating, we need to encourage them to self-actualize and thereby access the higher stages of development where they become concerned about the universal good.

In *The Listening Society: A Metamodern Guide to Politics, Part One,*[47] Hanzi Freinacht expresses a similar idea:

> "Our current society is designed to achieve growth, and of industrial output and redistributing its spoils. Future society must expand upon today's society's way of functioning; its institutions must be geared towards achieving more psychological goals. More goals of the soul."

As long as people are operating from the ego stages of development, they will be using the first algorithm or evolutionary intelligence to meet their needs. This algorithm enables people to become stronger and more resilient by focusing on themselves. In the early stages of life—the childhood and teenage years—while we are learning how to get our needs met in the

---

[47] Hanzi Freinacht, *The Listening Society: A Guide to Meta-modern Politics, Part 1* (Metamoderna ApS), 2017.

culture of our upbringing, this is an appropriate strategy. However, if we continue using this strategy to solve our problems once we have reached maturity, this strategy will prevent us forming cohesive human group structures. We will be focused on our own self-interest rather than the interest of the groups with which we identify.

Only when we transcend the individuating stage of psychological development are we mature enough to access the second algorithm of evolutionary intelligence. This algorithm enables people to become stronger and more resilient by bonding with similar others to form cohesive group structures—our self-interest becomes tied to the interest of the group with which we identify: we can only get our needs met if our group gets its needs met. We begin to focus on the common good.

Only when we move beyond the individuating and self-actualizing stages of development can we access the third algorithm of evolutionary intelligence. This algorithm enables groups of people to become stronger and more resilient by cooperating with other groups of similar or dissimilar others to form higher order group structures. We can see the expression of this algorithm attempting to operate in interdependent group structures such as the United Nations and the European Union. However, these two higher order group structures are currently struggling because there are not enough people in the member nations operating from the integrating and serving stages of development. In fact, most of the nations in the United Nations and several of the member nations of the European Union do not have a critical mass of people operating at the individuating stage of development. Consequently, there are ego-driven factions in many of these nations that put their needs ahead of their nation's needs. These factions find it very difficult to bond with dissimilar others in their own countries and in the higher order group structures of the United Nations and European Union.[48]

Where the third algorithm worked well was during World War II, when the Allied Nations opposed the expansionist activities of the Axis powers—Germany, Italy, and Japan. At the start of World War II, the Allies consisted of France, Poland, and the UK and the dominions of the British Commonwealth—Australia, Canada, New Zealand, and South Africa. Shortly thereafter, the Netherlands, Belgium, Greece, and Yugoslavia

---

[48] I cover this topic in detail in my book *Worldview Dynamics,* to be published a few months after this book.

joined the Allies. Later, the USSR and the US also joined. This alliance was formalized on the 1st of January 1942. Once the war was over, and the Allies had won, this temporary higher order group structure morphed into the higher order group structure known as the United Nations.

Something similar happened in North America in the eighteenth century. The colonization of the land that is now known as the United States of America began when immigrants from England settled in the colony of Virginia in 1607, in a town called Jamestown. By the late eighteenth century there were thirteen colonies, each with their own governments and monetary systems. Although all the colonies were governed by the British Parliament, they were, to all intents and purposes, sovereign states with their own constitutions, their own state laws, and their own currencies. In 1775, the thirteen colonies cooperated with one another to fight a common external threat—the British Parliament. The core issue was *no taxation without representation*. Every colony wanted freedom from the meddling of the British government in their monetary affairs.

They formed a temporary alliance and cooperated to fight the army of the British. When the war was over in 1776, they made the alliance permanent. They drafted the US constitution and established a permanent higher order entity we now know as the United States. In this example, we see people bonding together to form colonies—the second algorithm of evolutionary intelligence—and then colonies cooperating with one another according to the third algorithm of evolutionary intelligence to create a higher order entity—the United States of America.[49]

The main lesson we can draw from these two examples is that survival is, as I explained earlier, the primary trigger of evolutionary intelligence. It enables us to become individually stronger; it enables us to bond to form resilient group structures; and it enables group structures to cooperate to create a higher order entity.

Does this mean we need a new global survival crisis to elevate the consciousness of the human species—to get all the nations operating from the third algorithm of evolutionary intelligence? I am afraid I think it does. What will that crisis be? In my mind, the answer to this question is climate change (global warming). Only when we truly begin to address this issue

---

[49] In my forthcoming book *Worldview Dynamics and the Consciousness of Nations* I will explore in detail the role played by the algorithms of evolutionary intelligence in the creation of the European Union.

will we see all the nations of the world cooperating with one another—using the third algorithm of evolutionary intelligence to create a world in which the concept of humanity is palpable. Unfortunately, if we do not act soon, we may be too late.

What is the alternative?

There is an alternative if we want to avoid this survival crisis. The alternative is to support individuals in every nation in meeting their deficiency needs and creating an education program, customized for each nation, that supports people in differentiating and learning how to individuate and self-actualize. In my mind, such a program is a fundamental necessity if we are going to meet the UN Sustainable Development Goals.

Such a program would begin with school age children by giving them a values-based education, teaching them how to meditate, how to express their feelings and how to access their gifts and talents, how to connect with others in unconditional loving relationships, and how to contribute to the well-being of others. It would then focus on supporting teenagers and young adults in traversing the differentiating and individuating stages of psychological development, so they can focus on their growth needs.

IMPORTANT TERMS DEFINED

Algorithm

An algorithm is an established process or set of rules to be followed for problem solving. The concept of algorithm has existed for centuries. Nowadays, it is mainly used in data processing.

Attachment theory

Attachment theory is a psychological model which attempts to explain how humans respond when separated from loved ones. Attachment begins early in life, usually with the mother. It is important for the future social and emotional development of our species for everyone to experience a sense of attachment at a young age.

Beliefs

Beliefs are always personal. Some of them may originate at the cultural level, but they always belong to the individual. Beliefs are assumptions we hold to be true. They may or may not be true; that is why I call them assumptions.

Beliefs are always contextual. Our principal beliefs are formed in the specific physical, social, and cultural frameworks of existence that we are brought up in during the first twenty-four years of our lives, when our minds and brains are growing and developing—when we are learning how to get our deficiency needs met. Beliefs can be unconscious, subconscious, or conscious.

Unconscious beliefs

These are the beliefs (I also call them imprints) we learn while the reptilian mind/brain is growing and developing. The reptilian mind/brain is dominant—the main interface between "self" and the external world and the executive decision-making authority—from the first trimester of gestation up to around the age of two. Our personal imprints are triggered whenever we experience an event that reminds us of a trauma or a time when we struggled to stay alive and feel nutrutred during the period when the reptilian mind/brain was growing and developing.

Subconscious beliefs

These are the beliefs we learn while the limbic mind/brain is growing and developing. The limbic mind/brain is dominant from around the age of two until the age of about seven. Our subconscious beliefs are triggered whenever we experience an event that reminds of us of a situation that occurred while the limbic mind/brain was growing and developing—positive experiences when we got our safety needs met and negative experiences when we struggled and failed to get our safety needs met.

Conscious beliefs

These are the beliefs we learn while the neocortex mind/brain is growing and developing. The neocortex mind/brain is dominant from around the age of eight onwards. Our conscious beliefs are triggered whenever we experience changes in our lives. We use our beliefs to understand what is happening so that we can decide whether the situation we are experiencing is a threat to our safety or security or an opportunity to increase our safety or security.

DNA

Viewed through the lens of 3-D material awareness, deoxyribonucleic acid (DNA) is a molecule that carries the genetic instructions used in the development, functioning, and reproduction of living organisms. Viewed through the lens of 4-D energetic awareness, DNA is an energetic field of information and instructions that belong in part to the energy field of the species and in part to the energy field of the soul templates of the parents that made a child. In 3-D material awareness, specific parts of a DNA molecule are called genes. In 4-D energetic awareness, genes are specific aspects of the energy field of a DNA molecule.

Ego

Your ego is a field of conscious awareness that identifies with your physical body. Because the ego believes it inhabits a body and lives in a material world, it lives in three-dimensional material reality and thinks it can die. Because it thinks it can die, it thinks it has needs; and because it thinks it has needs, it develops fears about not being able to get its needs met. The principal needs of the ego are survival, safety, and security. The ego mind is the creation of the soul mind. The soul creates the ego to protect itself from the pain (energetic instability) of separation it experiences from being present in three-dimensional material awareness.

The ego is not *who* you are; it is who you *think* you are. It is the mask you wear to get your needs met in the physical, social, and cultural framework of your material existence. The ego represents your sense of identity in relation to others and the social context in which you live. Your ego identity begins to form during the first two to three years of your life, and, if all goes well, it reaches a natural resolution during your early twenties as you become a viable and independent member of your community in the cultural framework of your existence.

When you get to this stage in your life, you normally respond to the question "Who am I?" by stating your age, gender, role/occupation, race, religion, and nationality. These are the things that define your ego identity.

When the content and memories of the ego mind are known to us in our present-moment awareness, they are said to be conscious. When the content and memories of the ego mind are not known to us in our present-moment awareness, they are said to be either subconscious or unconscious. Subconscious content and memories can be easily brought into the conscious awareness of the ego mind. Unconscious content and memories fall into two categories: those that can be brought into the conscious awareness through specific, directed psychotherapeutic interventions, and those that cannot be brought into conscious awareness.

In the former category are the traumas we experience *in utero* and during the first two years of our lives when the reptilian mind/brain is the dominant decision-making authority. In the latter category is the content and memories of the reptilian mind/brain that control the homeostatic (biological) functions that keep us alive. We only become conscious of our homeostatic functions when the body's survival needs are not being met—whenever we experience uncomfortable physical sensations or physiological pain.

Epigenetics

Epigenetics is the study of heritable characteristics that do not involve alterations in DNA sequencing. They are features that lie on top of traditional genetic bases of inheritance. Epigenetics is most often used to describe changes that affect gene activity and expression. These changes are seen in observable characteristics and behaviour.

Instincts

Instincts are the beliefs of the body mind that exist at the level of the species mind. They are designed to meet the body's survival needs. An organism does not need a brain to have instincts, but it does need a mind. Cells, for example, do not have a brain, but they have instinctual responses that enable them to react to threats, protect themselves, and find energy sources that enable them to survive. Wherever there is awareness, there is mind; and the mind is always in the energy field.

Minds and brains

I frequently juxtapose the words "mind" and "brain" together to form the expression "mind/brain." I use this form of expression to emphasize that the mind and the brain are not the same. They exist in different realms of reality. The reptilian brain is the 3-D material instrument of the 4-D energetic mind that coordinates the functioning of the body. Causation starts in the 4-D energetic mind and is translated into physical action by the 3-D material brain.

When I want to emphasize either the data gathering and information processing aspect of the mind/brain, I use the term "brain." When I want to emphasize the meaning-making and decision-making mechanism of the mind/brain, I use the term "mind." Since everything that is composed of matter has an energy field, I also, from time to time, refer to the "mind" as the "energy field of the brain."

Once again, to give emphasis to the different realms of operation of the mind and brain, I sometimes use the term "body mind" to refer to the meaning-making and decision-making aspect of the reptilian mind/brain, and I use the term "emotional mind" to refer to the meaning-making and decision-making aspect of the limbic mind/brain. I also use the term "rational mind" to refer to the meaning-making and decision-making aspect of the neocortex mind/brain. Normally, when I refer to the ego mind, I am referring to the emotional mind and rational mind as a single unit to differentiate it from the body mind and soul mind.

Personal conscious

The personal conscious is the executive decision-making centre that interprets and gives meaning to what is happening in our world. It contains our conscious memories. We use our personal conscious to make rational or logical decisions about how to respond to changes in our external environment so that we can get our needs met. As adults, the processes, content, and memories of the personal conscious are related to the operation of the neocortex mind/brain.

Personal subconscious

The personal subconscious supports the conscious mind in making decisions when the personal conscious is engaged in thinking about other matters. The personal subconscious contains memories that, although not immediately accessible, can be brought into conscious awareness. The content and memories held in the personal subconscious can affect how we react to situations when a present-moment experience triggers an emotional memory stored in the subconscious mind. As adults, the processes, content, and memories of the personal subconscious are related to the operation of the limbic mind/brain, which I also refer to as the subconscious ego mind.

Personal unconscious

The personal unconscious supports the conscious mind in making decisions about the regulation and functioning of the body. The personal unconscious contains memories or imprints that are not readily available for inspection. It requires special therapeutic skills or hypnotherapy to access this content. The content and memories held in the personal unconscious affect not only our moods, behaviour, and decision-making, they also affect our physical and mental health. The processes, content, and memories of the personal unconscious are related to the operation of the reptilian mind/brain which I also refer to in this book as the body mind.

All living creatures have some form of body mind that controls the body's functioning. The awareness and processes that control the body mind are not accessible to the conscious mind. The body mind communicates with the personal conscious through physical sensations, bodily discomfort, and physiological pain.

Psyche

The term "psyche" has a long history of use, dating back to ancient times. In the days of the ancient Greek civilisation (800 B.C.–146 B.C.), the term "psyche" was used to refer to the soul. The idea of the soul was central to the philosophy of Plato. He considered the psyche to be immortal. In modern times, the meaning of the term "psyche" has been subject to significant change.

The internationally renowned psychiatrist Carl Jung makes a distinction between psyche and soul. He refers to the psyche as the totality of all human mental processes, conscious as well as unconscious. He refers to the soul as a clearly demarked functional complex that has a unique personality. For Jung, the soul was an aspect of the psyche, and, therefore, Jung's approach to psychoanalysis included not just the ego aspects of our personality but also the soul aspects. For Jung, the soul was part of our unconscious.

In recent decades, the terms "psyche" and "soul" have fallen into disuse, mainly because modern science finds it difficult to deal with what cannot be perceived by the senses or what originates from a domain that is regarded as the unknown. What is not known, and cannot be scientifically proven, does not exist as far as modern psychologists are concerned. Unconscious content that percolates into conscious awareness and synchronistic experiences are not given the attention they deserve. These impulses originating in the four-dimensional energetic world of the soul are mostly disregarded.

In this book, I align myself with the position of Carl Jung. I use the term "psyche" to refer to the totality of our mental processes (conscious, subconscious, and unconscious), and I use the term "soul" to refer to the functional complex that contains our own unique, life-transcending personality—the source of our four-dimensional awareness.

Rapid emergent learning

I call the learning that takes place during the first twenty-four years of our lives "rapid emergent learning" because we are learning at the same time as our mind/brains are forming. Consequently, the imprints and beliefs we learn during this period of our lives tend to become hard-wired into our brains in the form of synaptic connections. What we learn, particularly in the womb and during the first two years of life when the reptilian mind/brain (body mind) is growing and developing and the next five years when the limbic mind/brain (emotional mind) is growing and developing, conditions our reactions and responses to life events for the rest of our lives. Normal emergent learning takes place after our rational mind/brain has become fully functional around our mid-twenties.

Soul

Your soul is a field of conscious awareness that identifies with your energy field. It is who you are. You don't *have* a soul, you *are* a soul. Your soul and the soul of every other human being is an individuated aspect of the universal energy field from which everything in our physical world arises. Because the soul identifies with your energy field and not with your physical body, your soul lives in four-dimensional energetic reality. The soul knows it cannot die and consequently has no fears. Not only does the soul have no fears, it also has no needs. The reason it has no needs is that at the energetic level of its existence, it creates what it desires through its thoughts.

Because our souls are individuated aspects of the universal energy field, they feel a sense of connectedness to every other soul. Consequently, at the soul level, we live in oneness. There is no separation. When you live in a world of oneness, giving is the same as receiving: when you give to others, you give to yourself.

Even though the soul has no needs in the way that the ego has needs, it does have desires. The soul's principal desires are self-expression, connection, and contribution. The soul incarnates into three-dimensional material awareness to fulfil these desires. The purpose of the soul's desires is to recreate its four-dimensional energetic reality in three-dimensional material awareness. You know your soul's desires are being met when you feel your life has meaning, when you can connect with others at a deep level, and when you can use your gifts and talents to contribute to making a difference in the world. The only things preventing the soul from fulfilling its desires are the ego's fears about meeting its deficiency needs—the ego's survival, safety, and security needs. The ego's fears about meeting its deficiency needs keep it firmly attached to its physical, social, and cultural identity.

The soul incarnates by willing itself to be present in three-dimensional material reality. The soul's will to be present in three-dimensional material reality is the source of the will of the body mind to stay alive and the will of the ego mind to survive.

State, nation, country, and nation-state

A State (the S is always capitalised) is a political unit that has a sovereignty over the territory and the people in it. Sovereignty means ultimate, legitimate political authority. A State may be comprised of states (lower case "s").

A nation need not have physical boundaries. It is more about people bonded by culture and/or religion. For example, Palestine is a nation—there are people living in many geographically dispersed areas around the world who identify themselves as Palestinians, though no such State exists.

A country is just another word for State and can be used interchangeably. However, since "country" has other meanings, State is a more precise term to use in a political discourse. A nation-state is a country in which a distinct cultural or ethnic group inhabits a territory and has formed a State.

Triune brain

The reptilian brain, the limbic brain, and the neocortex are together known as the triune brain. The triune brain is a model of the evolution of the brain proposed by the American physician and neuroscientist Paul D. MacLean. MacLean originally formulated his model in the 1960s and propounded it at length in his book *The Triune Brain in Evolution.*[50] The triune brain consists of the reptilian complex, the paleomammalian complex (limbic system), and the neo-mammalian complex (neocortex). Each of these brain structures are viewed as being sequentially added during the process of evolution.

Universal energy field

The universal energy field, sometimes called the Great Field or the Zero-Point Energy Field, is the ground of all being from which everything in our material world arises. The origin of this ground of all being is referred to by the scientific establishment as the "Big Bang," a cosmological event that occurred 13.8 billion years ago that gave rise to our physical universe.

The universal energy field is the energetic container for everything that exists. It is a field of universal conscious awareness and intelligence that forms the background from which everything in our physical world emerges. The universe was born as an undifferentiated unity.[51]

---

[50] Paul D. McLean, *The Triune Brain in Evolution* (New York: Plenum Publishing), 1990.

[51] Gerald L. Schroeder, *The Hidden Face of God: Science Reveals the Ultimate Truth* (New York: Touchstone), 2001, p. 41.

INDEX

Numbers

2-D awareness 9, 11
3-D awareness 9, 10, 11, 13, 22
3-D material awareness 13, 18, 19, 20, 23, 39, 80, 81
4-D awareness 21
4-D energetic awareness 39, 80, 81

A

Abraham Maslow 6, 71
Adaptive mutation 18, 19
Adaptive thinking 19
Aksel Sandemose 73
Algorithm 4, 23, 24, 30, 31, 32, 33, 37, 41, 42, 43, 56, 57, 65, 66, 67, 68, 70, 75, 76, 77, 78, 79
Atomic plane xvii, xviii, 25
Attachment Theory 46, 79

B

Big Bang xvii, 2, 7, 8, 23, 24, 87
Bildung 70, 71
BREXIT 68

C

Carbon atom xviii, 25, 26, 27, 28, 30
Cellular plane xvii, xviii, 25
Charles Robert Darwin 15

Comb analogy 12
Conforming 38, 40, 43, 47, 48, 53, 58, 60, 61, 70
Consciousness of nations 66, 77
Creationism 16

D

Decision-making 4, 8, 20, 31, 32, 40, 47, 67, 70, 80, 82, 83, 84
The desires of the soul 5
Differentiating 38, 40, 43, 48, 51, 54, 58, 61, 70, 71, 78
DNA 14, 16, 19, 20, 21, 22, 23, 45, 80, 81, 82
DNA as coding 21

E

Edward O. Wilson 12
Ego development 38
Ego-soul alignment 41
Ego-soul dynamics 37, 38
Embryo 22, 39, 44, 45
Energetic instability 21, 23, 27, 81
Energetic stability 3, 5, 6, 21, 24, 27, 30, 31
Epigenetics 45, 82
Eukaryotic cell xviii, 25, 27, 28, 30
European Union 25, 76, 77
Evolution 1.0 15, 17

S

T

U

W

www.ingramcontent.com/pod-product-compliance
Lightning Source LLC
Chambersburg PA
CBHW022100170526

45157CB00004B/1411